MATHEMATICS
FOR
BUSINESS

Seventh Edition

MATHEMATICS

FOR

BUSINESS

Seventh Edition

Gary Bronson, Ph.D.
Fairleigh Dickinson University

Richard Bronson, Ph.D.
Fairleigh Dickinson University (Emeritus)

Maureen Kieff
Fairleigh Dickinson University

MERCURY LEARNING AND INFORMATION
Dulles, Virginia
Boston, Massachusetts
New Delhi

Publisher: David Pallai
MERCURY LEARNING AND INFORMATION
22841 Quicksilver Drive
Dulles, VA 20166
info@merclearning.com
www.merclearning.com
1-800-232-0223

G. Bronson, R. Bronson, M. Kieff. *Mathematics for Business, Seventh Edition.*
ISBN: 978-1-68392-766-2

The publisher recognizes and respects all marks used by companies, manufacturers, and developers as a means to distinguish their products. All brand names and product names mentioned in this book are trademarks or service marks of their respective companies. Any omission or misuse (of any kind) of service marks or trademarks, etc. is not an attempt to infringe on the property of others.

Library of Congress Control Number: 2021943215

212223321   Printed on acid-free paper in the United States of America.

Our titles are available for adoption, license, or bulk purchase by institutions, corporations, etc. For additional information, please contact the Customer Service Dept. at 800-232-0223(toll free).

To Oliver and Benjamin
—Gary Bronson

To Casley, Sarah, Asher, and Max
—Richard Bronson

To Richard, Kristen, Annie, and My Parents
—Maureen Kieff

CONTENTS

*P*REFACE

Quantitative methods have become essential in economic forecasting, allocation of resources, portfolio analysis, inventory analysis, data-mining, and innovative solutions to myriad new social and climate challenges. The aim of this text is to introduce and provide a basic and necessary understanding of these quantitative methods in a student centered and affordable text.

We understand that the average student knows that mathematics is a powerful tool for solving problems, but may feel uneasy, sometimes fearful, about the subject matter. Our primary concern in writing this text is to present the material in a clear, understandable, and non-intimidating manner. By numerous examples each new mathematical concept is tied to a practical problem, so that the student never loses sight of the ultimate goal: to develop mathematical tools to solve business and social problems. We are not interested in mathematics as an end unto itself. Additionally, because many of the topics are now solved using spreadsheets, we show where, when, and how to apply these increasingly necessary tools.

The text assumes student familiarity with algebraic concepts but not facility in using them. As such, Chapter 1 is a review of elementary concepts typically covered at the secondary school level.

Chapters 2 and 3 deal with mathematical equations, and graphs. Material on polynomial equations, their solutions, and graphs is presented.

Interest rates, cash flows, and annuities are presented in Chapter 4. This serves as both motivation and introduction to applying mathematical equations to areas typically of interest to students. Here, mortgages, consumer loans, annuities and their relationship to pension plans and lottery winnings paid over time are presented. The subject matter is relatively easy to grasp and since most students find it interesting, it provides an opportunity for early success in using mathematical methods.

The next two chapters (5 and 6) bring the student to an appreciation and awareness of differential calculus as a powerful mathematical approach for analyzing and modeling commercial systems. The applications in these chapters were developed with this philosophy in mind. They provide a setting for introducing mathematical models and exposing students to realistic applications of average and instantaneous rates of change. Although we do not expect the reader of the book to become an expert in calculus, we do hope that these applications develop an understanding of how differential calculus is used in real-life problems.

The material in Chapter 7, on least-squares analysis, is included to answer the usual question asked by students as to the origin of the equations they have been using. An integral part of this chapter is the use of spreadsheets to create trend lines and their associated equations.

A number of people have been instrumental for helping to make this book a reality. First, our appreciation goes to our students, who used most of the material in this book in prepublication form. We sincerely hope this text provides a very low-cost alternative to the otherwise extremely expensive texts they have been required to purchase in the past, with the same quality that has characterized all of our published texts. We would also like to express our thanks to the hard-working and dedicated staff at Mercury Learning and Information. To David Pallai, Publisher, for expressing confidence in and publishing this book, and for tirelessly working with us closely on all aspects of the publication process, from beginning to end. It is rare that authors can receive the "personal touch" from such a dedicated and experienced publishing professional, who guided us through the various intricacies and challenges of the process. To Jennifer Blaney, for her professionalism, production expertise, and in doing a great job keeping the book on schedule. In particular, we would like to recognize the commitment to quality that has been shown by these individuals, in terms of the care shown during copyediting, resolving issues, requesting multiple rounds of page proofs, and providing us the opportunity to provide input into just about every aspect of our book and its production, marketing, and promotions. For this, we are grateful. Finally, and most importantly, we owe our deep appreciation and thanks to our spouses, Rochelle, Evelyn, and Richard.

Gary Bronson, Ph.D.
Richard Bronson, Ph.D.
Maureen Kieff
September 2021

ABOUT THE AUTHORS

Gary Bronson, Ph.D., is a professor of marketing, information systems, and decision sciences at the Silberman College of Business, Fairleigh Dickinson University, where he was twice voted Teacher of the Year of the college and received the Distinguished Research award of the university. He has worked as a senior engineer at Lockheed Electronics, an invited lecturer and consultant to Bell Laboratories, and a software consultant to a number of Wall Street financial firms. He is the author of the highly acclaimed *A First Book of C* and has authored several other successful programming textbooks on C++, Java, and Visual Basic. He is a co-author of *Excel Basics* and the *Excel 2019 Project Book* with Jeffrey Hsu. Additionally, he is the author of a number of journal articles in the fixed-income financial and programming areas. Dr. Bronson received his Ph.D. from Stevens Institute of Technology.

Richard Bronson, Ph.D., is Professor Emeritus in the Department of Mathematics at Fairleigh Dickinson University, where he has won the university's Distinguished Teaching, Research, and Faculty Service awards, including the Distinguished College or University Teaching award by the New Jersey Section of the Mathematical Association of America. He has authored several successful texts in matrix methods, differential equations, linear algebra, finite mathematics, and operations research. He has also published a number of journal articles in mathematics and system simulation. His latest publication is a political thriller titled, *Antispin*.

Maureen Kieff, is a Clinical Assistant Professor in the Department of Information Systems and Decision Sciences at Fairleigh Dickinson. She has received 12 awards during her career for outstanding teaching and service to students, the latest being the 2014 College Teacher of The Year. She has co-authored ten research articles published in academic journals, primarily addressing mutual fund performance.

THE BASICS

In this Chapter

This chapter is a review of the topics in algebra that are used throughout the text. Readers who already have a working knowledge of this material are advised to skim over the chapter and go directly to Chapter 2. Others are advised to spend as much time as is necessary to master this material before proceeding further.

1.1   SIGNED NUMBERS

Many people have no difficulty performing the basic arithmetic operations (addition, subtraction, multiplication, and division) on positive numbers, but find similar operations on negative numbers mind–boggling. These operations typically are much easier to understand as they relate to profit and loss in commercial settings

In business, among many other uses, positive numbers represent profits, and negative numbers represent losses. In particular, −$5.00 denotes a loss of $5.00, −$10.25 denotes a loss of $10.25, and +$3.00 denotes a profit or gain of

$3.00. By convention, a number without a sign is considered positive. Therefore, $3 = +3$, $5 = +5$, and $9.75 = +9.75$.

In the context of profits and losses, the addition sign is read "followed by." Then, $-3 + 5$ is a $3 loss *followed by* a $5 gain; the result is a net gain of $2, so $-3 + 5 = 2$. For some, the process is clarified when viewed in the context of betting at a horse race. If a person loses $3 on the first race and then wins $5 on the second race, he or she will have a net gain of $2. Again, $-3 + 5 = 2$.

To calculate $3 + (-7)$, we reason similarly. A $3 profit followed by a $7 loss results in a net loss of $4. Alternatively, if a person wins $3 on the first race but loses $7 on the second race, he or she will then be behind by $4. Either way, $3 + (-7) = -4$.

The same reasoning is valid when adding two negative numbers. The quantity $(-10) + (-8)$ denotes a $10 loss followed by an $8 loss. Or it can be viewed as a person losing $10 on the first bet and then losing $8 on the second bet. The end result is the same, a total loss of $18. Therefore $-10 + (-8) = -18$.

Viewed as profits and losses, the following results should be straight forward:

$$5 + 7 = 12$$
$$-9 + 3 = -6$$
$$2 + (-4) = -2$$
$$8 + (-5) = 3$$
$$-7 + (-8) = -15.$$

The multiplication of signed numbers is a two–step operation. The first step is to multiply the numbers disregarding any negative signs (treat all numbers as positive). The second step is to determine the appropriate sign for the result. Here the following rules apply:

Multiplication Rules:

- The product of two positive numbers is positive.
- The product of two negative numbers is positive.
- The product of two different signed numbers is negative.

Thus, when both numbers have the same sign (either both positive or both negative); the result is positive. When both numbers have different signs (one positive and one negative), the result is negative. For the multiplication of more than two numbers, the following rules apply:

- The product of all positive numbers is positive
- The product of an even number of negative numbers is positive.
- The product of an odd number of negative numbers is negative.

To calculate −5 times +2, first multiply 5 by 2 (disregarding all negative signs) and obtain 10. To determine the appropriate sign for −5 times 2, we note that both numbers have different signs (a negative 5 and a positive 2), so the result is negative. Accordingly −5 times + 2 is −10.

Mathematically, two sets of parentheses next to each other denote multiplication. We write (3)(7) for 3 times 7, (8)(−4) for 8 times −4, and (−3) (−6) for −3 times −6. In particular, (−5)(2) denotes −5 times 2, which was just found to be (−5)(2) = −10.

To calculate (8)(−½), first disregard the negative sign and multiply the positive numbers, obtaining (8)(½) = 4. Because +8 and −½ have different signs, their product is negative. Accordingly, (8)(−½) = −4.

To calculate (−3)(−6), first calculate (3)(6) = 18. Because −3 and−6 have the *same* sign (both negative), we conclude that (−3)(−6) = 18. Similarly,

$$(3)(-4) = -12$$
$$(-3)(4) = -12$$
$$(-3)(-4) = 12$$
$$(5.1)(-0.2) = -1.02$$
$$\left(-\frac{1}{3}\right)\left(-\frac{5}{2}\right) = \frac{5}{6}$$

Division follows the same pattern as multiplication. First, divide the two numbers disregarding signs and assuming all numbers to be positive. Then determine the appropriate sign of the result exactly as in multiplication. Thus, the following division rules apply:[1]

Division Rules:

- The quotient of a division problem is positive if both the numerator and denominator have the same signs.
- The quotient of a division problem is negative if the numerator and denominator have opposite signs.

Thus, to calculate −6 ÷ 3, for example, or equivalently, −6/3, first divide +6 by +3, obtaining 2. Because −6 and +3 have different signs, the quotient is negative. Thus, −6/3 = −2. To calculate −12/−3, we first find 12/3 = 4. Because both −12 and −3 have the same sign, their quotient is positive, and −12/−3 = +4. Similarly,

---

[1] Division by 0 is not defined and 0 divided by any non-zero number is 0. That is, for any real number $a \neq 0$, $a/0$ is not defined and $0/a = 0$.

$$8/-4 = -2$$
$$4/-8 = -0.5$$
$$-11/-5 = 2.2$$
$$2.1/-3.2 = -0.65625.$$

The most difficult operation for many is subtraction. For subtraction, the following rule applies:

For any real numbers a and b, a − b = a + (−b).

That is, subtraction can be converted to addition, and the standard addition rules will then apply. The key step is to introduce a plus sign before the subtraction sign and then incorporate the minus sign into the second number. We write $8 - 10$ as $8 + (-10)$; we write $-7 - 11$ as $-7 + (-11)$; we write $-3 - (-8)$ as $-3 + [-(-8)]$. Each subtraction then becomes an addition. If, as a result, there are two negative signs next to each other, such as $-(-8)$, the following rule applies:

The negative of a negative number is the number itself; that is, −(−a) = a.

Thus, the expression $-(-8) = +8$

To calculate $-7 - 11$, we rewrite the expression as $-7 + (-11)$. This can be considered as a $7 loss followed by a $11 loss, which results in a net, which results in a net loss of $18. That is,

$$-7 - 11 = -7 + (-11) = -18.$$

Similarly,

$$8 - 10 = 8 + (-10) = -2$$
$$-3 - (-8) = -3 + [- (-8)] = -3 + 8 = 5$$
$$-9 - (-2) = -9 + [-(-2)] = -9 + 2 = 7$$
$$7 - 4 = 7 + (-4) = 3$$
$$-3 - 4 = -3 + (-4) = -7$$

Occasionally, one is faced with a series of operations such as $(-8)(-3 + 5)$. The procedure is to combine two numbers at a time. Because $(-8)(-3 + 5)$ is -8 times the quantity $-3 + 5$, we first calculate the sum $-3 + 5$, which equals 2. Then,

$$(-8)(-3 + 5) = (-8)(2) = -16$$

A more complicated expression is $7[-3 - 2(8 - 10)]$. When calculating any expression having more than two numbers, the following order of operations must be followed:

Order of Operations:

1. Parenthesis – calculate all expressions within parentheses first, starting with the innermost parentheses, or brackets and then systematically working toward the outermost parentheses or brackets. If there is only one set of parentheses, calculate the expression within the parentheses before proceeding with the remaining operations.

2. Exponents – perform all exponents[2]

3. Perform all multiplications and divisions – left to right

4. Perform addition and subtraction – left to right

Thus, in the expression $7[-3 - 2\,(8 - 10)]$, we first compute $(8 - 10)$ within the innermost parentheses and then multiply this result by 2. Accordingly,

$$7[-3 - 2\,(8 - 10)] = 7[-3 - 2\,(-2)]$$
$$= 7[-3 + (-2)(-2)]$$
$$= 7[-3 + 4] = 7[1] = 7$$

Similarly,

$$5[3(2 - 8) - 4(5 - 6)] = 5[3(-6) - 4(-1)]$$
$$= 5[-18 + (-4)(-1)]$$
$$= 5[-18 + 4]$$
$$= 5[-14]$$
$$= -70$$

and

$$[5 + (-3)][2 - 7] = [2][2 - 7]$$
$$= 2[2 + (-7)]$$
$$= 2(-5)$$
$$= -10$$

All signed numbers obey specific properties of arithmetic. For example, the order in which addition is performed is immaterial. In particular, $5 + 7 = 7 + 5$, $-2 + 8 = 8 + (-2)$, $5 + [8 + (-3)] = [5 + 8] + (-3)$, and $-6 + [-7 + 4] = [-6 + (-7)] + 4$. The only requirement is that each number in a sum be added once. If we let a, b, and c denote any signed numbers, either positive or negative, we can state these rules formally as

---

[2] Exponents are presented in Section 1.3.

$a + b = b + a$ (commutative law for addition)
$(a + b) + c = a + (b + c)$ (associative law for addition)

Similar rules hold for multiplication; the order in which multiplication is performed is irrelevant. Clearly, $(5)(7) = (7)(5)$, $(-2)(8) = (8)(-2)$, $(5)[(8)(-3)] = [(5)(8)](-3)$, and $(-6)[(-7)(4)] = [(-6)(-7)](4)$. Again, the only requirement is that each number in a product be multiplied once. Formally,

$(a)(b) = (b)(a)$ (commutative law for multiplication)
$(a)[(b)(c)] = [(a)(b)](c)$ (associative law for multiplication)

The left side of the last equation indicates that first b and c are multiplied together, and the result is then multiplied by a. The right side indicates that first a and b are multiplied together, and the result is then multiplied by c. The equality indicates both procedures yield the same result. Specifically, with $a = -3$, $b = -2$, and $c = 8$, $(-3)[(-2)(8)] = (-3)(-16) = 48$ and $[(-3)(-2)](8) = [(6)](8) = 48$, which are indeed equal.

Other rules are useful if the operations of addition and multiplication are mixed. They are

$(a)(b + c) = (a)(b) + (a)(c)$ (left distributive law)
$(b + c)(a) = (b)(a) + (c)(a)$ (right distributive law).

If, as an example, $a = -3$, $b = -2$, and $c = 8$, the left side of the equation for the left distributive law becomes $(-3)(-2 + 8) = (-3)(6) = -18$, while the right side becomes $(-3)(-2) + (-3)(8) = 6 + (-24) = -18$.

Note that each side of the equations for the distributive laws is arithmetically different. On the left side, two numbers are first added, and the result is then multiplied by a. On the right side, two pairs of numbers are first multiplied, and the results are then added. The final results on both sides, however, are equal.

Exercises 1.1

Evaluate the following expressions.

1. $3 + (-6)$	**2.** $-4 + 7$	**3.** $19.7 + (-18.1)$
4. $-6.2 +) + (-8.1)$	**5.** $-9 + (-1/2)$	**6.** $-4.1 + 7$
7. $9(18)$	**8.** $9(-8)$	**9.** $(-9)(18)$
10. $(-9)(-18)$	**11.** $(2)(-1/3)$	**12.** $(-5)(-1/6)$
13. $(-6.1)(2.3)$	**14.** $(-8)(-1.4)$	**15.** $(-8)/(-2)$

16. $8/(-2)$ **17.** $-8/2$ **18.** $-2/8$

19. $4/(-5)$ **20.** $(-5)/(-4)$ **21.** $-22/4$

22. $8 - 4$ **23.** $4 - 8$ **24.** $-4 - 8$

25. $-4 - (-8)$ **26.** $-8 - 4$ **27.** $-8 - (-4)$

28. $2.1 - 5.6$ **29.** $-5.6 - 2.1$ **30.** $1/10 - 1/5$

31. $2[5 + (-3)]$ **32.** $-2[1 + (-6)]$ **33.** $-4(1 - 3) + 2(2 - 5)$

34. $6[2(-1 + 7) - 3]$ **35.** $(1.6)(1.9 - 2.1)$ **36.** $4[(-1)(2-9) + 7(3)]$

37. $\dfrac{8\left[1-(-8)\right]-2\left[7-1\right]}{2}$ **38.** $\dfrac{\left[(8-12)/4\right]+4(9-3)}{5-9}$

39. $\dfrac{(5-11)(8-14)+42(2+3)}{7\left[2(1+30)-3(2-5)\right]}$ **40.** $\dfrac{8\left[-3(4-1)-6(4-8)\right]}{2\left[-8-(-7)\right]}$

41. $\dfrac{9\left[3(2)-5(-1)\right]-6\left[(11-4)\right]}{5}$ **42.** $\dfrac{3\left[-2(7-4)-9(11-8)\right]}{6\left[5(4-1)+10(2-8)\right]}$

1.2   EXPONENTS

Exponents provide a convenient notation for representing the product of a number times itself many times. For example, consider the following, which are valid for any signed number a (either positive or negative)

$$a^2 = (a)(a)$$
$$a^3 = (a)(a)(a)$$
$$a^4 = (a)(a)(a)\,(a)$$

The definition of a number, denoted as a, raised to the nth power, where n denotes a nonnegative integer (whole number), is given by

$$a^n = (a)(a)(a)(a)\ldots\ldots(a) \leftarrow n \text{ values of } a \text{ multiplied together}$$

The quantity a^n is typically read as either "the nth power of a," or "a to the nth."

For example,

$$5^2 = (5)(5) = 25 \text{ is read as "5 squared is 5 times}$$
$$5 \text{ equals 25"}$$
$$(-4)^3 = (-4)(-4)(-4) = -64 \text{ is read as "-4 cubed is -4 times}$$
$$-4 \text{ times } -4 \text{ equals } -64\text{"}$$
$$(-1/3)^4 = (-1/3)(-1/3)(-1/3)(-1/3) = 1/81 \text{ is read as 1/3 to the 4}^{\text{th}} \text{ power}$$
$$\text{equals } 1/81\text{")}$$

and

$$2^{10} = (2)(2)(2)(2)(2)(2)(2)(2)(2)(2) = 1024 \text{ is read as "2 to the 10}^{\text{th}} \text{ power}$$
$$\text{equals } 1024\text{")}$$

Notice that it is much easier to write 2^{10}, than list 2 ten times. Additionally, as most calculators have an exponential function (usually with keys having either a y^x or $\wedge$ notation) it is easier to calculate the final numerical value using the designated exponential key than entering and multiplying the given number n times.

One consequence of the exponential definition is the property

$$(a^n)(a^m) = a^{n+m} \qquad \text{(Eq. 1.1)}$$

where n and m are positive integers. For example,

$$(6^2)(6^3) = [(6)(6)][(6)(6)(6)] = (6)(6)(6)(6)(6) = 6^5 = 6^{2+3}$$
$$(-2)^4(-2)^3 = [(-2)(-2)(-2)(-2)][(-2)(-2)(-2)] = (-2)^7 = (-2)^{4+3}$$

and

$$\left(-\frac{1}{3}\right)^4\left(-\frac{1}{3}\right)^2 = \left[\left(-\frac{1}{3}\right)\left(-\frac{1}{3}\right)\left(-\frac{1}{3}\right)\left(-\frac{1}{3}\right)\right]\left[\left(-\frac{1}{3}\right)\left(-\frac{1}{3}\right)\right] = \left(-\frac{1}{3}\right)^6 = \left(-\frac{1}{3}\right)^{4+2}$$

Equation 1.1 is valid only if the left side of the equation is a number raised to a power times that *same number* raised to a power. The formula is not valid if one a in Equation 1.1 is replaced by another number b. For example, Equation 1.1 is *not* applicable to the product $(2)^5(3)^4$.

A second useful property of powers is

$$(a^n)^m = a^{nm} \qquad \text{(Eq. 1.2)}$$

That is, any number a raised to a power n, which is itself raised to a power m, is equal to a raised to the power n times m. For example, note that

$$\left(2^2\right)^3 = \left(2^2\right)\left(2^2\right)\left(2^2\right) = [(2)(2)][(2)(2)][(2)(2)] = 2^6 = 2^{(2)(3)}$$

and

$$\left[\left(-\frac{1}{3}\right)^3\right]^3 = \left[\left(-\frac{1}{3}\right)\left(-\frac{1}{3}\right)\left(-\frac{1}{3}\right)\right]^3$$

$$= \left[\left(-\frac{1}{3}\right)\left(-\frac{1}{3}\right)\left(-\frac{1}{3}\right)\right]\left[\left(-\frac{1}{3}\right)\left(-\frac{1}{3}\right)\left(-\frac{1}{3}\right)\right]\left[\left(-\frac{1}{3}\right)\left(-\frac{1}{3}\right)\left(-\frac{1}{3}\right)\right]$$

$$= \left(-\frac{1}{3}\right)^9 = \left(-\frac{1}{3}\right)^{(3)(3)}$$

Using Equations 1.1 and 1.2 together, we have

$$\left(2^4\right)^3 \left(2^2\right)^4 = 2^{12} 2^8 = 2^{12+8} = 2^{20}$$

and

$$[(-4)^3]^5[(-4)^2]^3 = (-4)^{15}(-4)^6 = (-4)^{21}$$

Be careful not to confuse these two properties. A common error is to write $(a^n)^m = a^{n+m}$ or $(a^n)(a^m) = a^{nm}$. Both of these are *incorrect and lead to wrong answers*.

Equations 1.1 and 1.2 can be extended to negative powers if we first give meaning to what a negative exponent means. Accordingly, for any non-zero signed number a and any positive integer, we define

$$a^{-n} = \frac{1}{a^n} \qquad \text{(Eq. 1.3)}$$

Therefore,

$$5^{-2} = \frac{1}{5^2} = \frac{1}{25}$$

$$(4)^{-3} = \frac{1}{(4)^3} = \frac{1}{64}$$

$$(-4)^{-3} = \frac{1}{(-4)^3} = \frac{1}{-64} = -\frac{1}{64}$$

$$\left(-\frac{1}{3}\right)^{-4} = \frac{1}{\left(-\frac{1}{3}\right)^4} = \frac{1}{\left(\frac{1}{81}\right)} = 81$$

and

$$2^{-10} = \frac{1}{2^{10}} = \frac{1}{1024}.$$

It follows from Equations 1.1 and 1.2 that the expression, $2^7 2^{-3} = 2^{7+(-3)} = 2^4$, and that $(4^{-6})^{-5}$, for example, equals $4^{(-6)\,(-5)} = 4^{30}$.

It is also useful to define

$$a^0 = 1 \qquad\qquad\qquad \text{(Eq. 1.4)}$$

for every non-zero signed number a.[3] Accordingly, $5^0 = 1$, $(-4)^0 = 1$, and $(-5/8)^0 = 1$.

Using Equations 1.1 and 1.3, we have

$$\frac{a^n}{a^m} = a^n \left[\frac{1}{a^m} \right] = a^n a^{-m} = a^{n+(-m)} = a^{n-m}$$

Therefore,

$$\frac{a^n}{a^m} = a^{n-m} \qquad\qquad\qquad \text{(Eq. 1.5)}$$

Equation 1.5 is applicable for every real number a and all integers n and m. Thus, for example, using Equation 1.5,

$$\frac{5^7}{5^4} = 5^{(7-4)} = 5^3 = 125$$

$$\frac{2^6}{2^8} = 2^{(6-8)} = 2^{-2} = \frac{1}{2^2} = \frac{1}{4}$$

and

$$\frac{\left(-\dfrac{1}{3}\right)^2}{\left(-\dfrac{1}{3}\right)^5} = \left(-\dfrac{1}{3}\right)^{2-5} = \left(-\dfrac{1}{3}\right)^{-3} = \frac{1}{\left(-\dfrac{1}{3}\right)^3} = \frac{1}{-\left(\dfrac{1}{27}\right)} = -27$$

Equations 1.1 through 1.5 can be used to simplify tedious multiplication and division operations when each factor can be expressed as *the same number* raised to a power. For example, if one recognizes that $8 = 2^3$, $512 = 2^9$, $64 = 2^6$, and $1024 = 2^{10}$, then

---

[3] Note that 0^0 is not defined as 1, as it would lead to contradictions in more advanced mathematics.

$$\frac{8(512)}{(64)(1024)} = \frac{2^3 2^9}{2^6 2^{10}} = \frac{2^{3+9}}{2^{6+10}} = \frac{2^{12}}{2^{16}} = 2^{(12-16)} = 2^{-4} = \frac{1}{2^4} = \frac{1}{16}$$

Although Equations 1.1 through 1.5 were defined for any signed number a and all integers (whole numbers) n and m, they actually are also valid for non-integer exponents. To understand why this is so, we first have to give meaning to fractional exponents such as $\frac{1}{2}$, $\frac{1}{3}$, and $\frac{1}{10}$ in expressions such as $4^{1/2}$, $8^{1/3}$, and $120^{1/10}$. This is accomplished by defining the exponent $1/n$ as the nth root of the number a, which is frequently written as $\sqrt[n]{a}$.[4]

Calculators that have exponential capabilities can also calculate fractional exponents if the fraction is first converted to a decimal value; the decimal value is then used as the exponent.

Numbers raised to more complicated fractional exponents, such as $4^{7/2}$ and $27^{5/3}$, also can be defined mathematically, which we do now for completeness, noting beforehand that such numbers have very limited applications to business problems. In general, $a^{p/q} = (a^{1/q})^p$ when a is a positive number and both p and q are positive integers. Therefore, $4^{3/2} = (4^{1/2})^3 = 2^3 = 8$. Note that $a^{p/q}$ can also be defined as $(a^p)^{1/q}$. Using this definition, $4^{3/2} = (4^3)^{1/2} = (64)^{1/2} = 8$, which is, of course, the same result as obtained using the first definition. Using a calculator for fractional exponents is accomplished in the same manner as integer exponents. First, calculate the decimal value of the exponent, and then use this value as the exponent using the calculator's exponentiation capability.

Table 1.1 summarizes the exponential rules for *all* positive real numbers a, n, and m. Additionally, if both n and m are integers, these properties are also true for negative values of a.

TABLE 1.1

Definition	Example
$a^n a^m = a^{n+m}$	$2^3 2^5 = 2^8$
$(a^n)^m = a^{nm}$	$(3^5)^4 = 3^{20}$
$a^{-n} = 1/a^n$	$4^{-2} = 1/4^2 = 1/16$
$a^0 = 1$ (for any $a \neq 0$)	$25^0 = 1$
$a^n/a^m = a^{n-m}$	$5^7/5^3 = 5^4$

---

[4] By convention, the second root of a number, which is its square root, is written as $\sqrt{a}$, rather than as $\sqrt[2]{a}$.

There is one last property of exponents that is useful; it involves the product of two different positive numbers raised to the *same* exponent:

$$a^n b^n = (ab)^n$$

(Eq. 1.6)

For example,

$$(5.2)^3(2)^3 = [(5.2)(2)]^3 = (10.4)^3 = 1,124.864$$
$$(4)^{3.1}(7)^{3.1} = (28)^{3.1} = 30,633.02583$$

and

$$(1.2)^{-3.4}(1.1)^{-3.4} = \left[(1.2)(1.1)\right]^{-3.4} = (1.32)^{-3.4} = 0.38908871$$

Like the other properties, Equation 1.6 is also valid for negative values of *a* and negative *b*.

We conclude this section with a warning. Most other properties of exponents that are ingeniously invented at times of stress, for example, during an examination, are usually not valid. In particular, as previously noted

$$a^n a^m \neq a^{nm}$$
$$(a^n)^m \neq a^{n+m}$$

Additionally,

$$a^n b^m \neq (ab)^{n+m}$$

The safest procedure is to understand and memorize the five properties provided in Table 1.1 and to *assume* that all other properties are not valid unless you can prove them.

Exercises 1.2

In Exercises 1 through 28 simplify each of the given expressions into one exponent. Make sure that each solution is given with a positive exponent.

1. $\dfrac{3^5 3^4}{3^2 3^3}$

2. $\dfrac{7^2 7^{-3} 7^4}{7^8 7^{-2}}$

3. $\dfrac{\pi^4 \left(\pi^2\right)^3}{\left(\pi^{-2}\right)^4 \pi^3}$

4. $\left[\left(-\dfrac{1}{2}\right)^2\right]^4 \left[\left(-\dfrac{1}{2}\right)^{-3}\right]^2 \left[\left(-\dfrac{1}{2}\right)^{-4}\right]^{-5}$

5. $\dfrac{(1.7)^{8.1}(1.7)^{-3.4}}{(1.7)^{-4.1}(1.7)^{3.7}}$

6. $\dfrac{x^3 \left(x^2\right)^4 \left(x^{-3}\right)^7}{\left(x^{-3}\right)^{-4} x^5}$

7. $\dfrac{\left(y^{-3}\right)^{-2} y^4 y^{-1}}{y^2 \left(y^3\right)^{-1}}$

8. $\dfrac{\left(x^4\right)^2 \left(1/x\right)^{-3}}{\left(1/x\right)^2}$

9. $\left\{\left[\left(3.1\right)^{-2}\right]^{-4}\right\}^3$

10. $\dfrac{2^5 2^3}{2^2 2}$

11. $\dfrac{5^7 5^{-3} 5^2}{5^4 5^{-2}}$

12. $\dfrac{x^{11} x^{-13} x}{x^{-7} x^2 x^3}$

13. $\dfrac{\left(2.1\right)^3 \left(2.1\right)^7}{\left(2.1\right)^8 \left(2.1\right)^{-2} \left(2.1\right)^{-5}}$

14. $\dfrac{\left(y^3\right)^{-2} \left(y^{-5}\right)^2 y^7}{\left(y^{-2}\right)^4 \left(y^{-2}\right)^{-4}}$

15. $\pi^{-3} \left(\pi^2\right)^7 \pi^{-8}$

16. $\left(\dfrac{1}{3}\right)^7 \cdot \left(\dfrac{1}{3}\right)^{-8} \cdot \left(\dfrac{1}{3}\right)^4$

Using a calculator, determine the values of the quantities given in Exercises 17 through 28.

17. $9^{3/2}$

18. $16^{-5/4}$

19. $27^{2/3}$

20. $100^{-3/2}$

21. $(3^{1/2})(12^{1/2})$

22. $(3^{1/3})(9^{1/3})$

23. $(5)^{-1/2}(20)^{-1/2}$

24. $(2^{-3/2})(32)^{-3/2}$

25. $\sqrt{9/4}$

26. $\sqrt{\dfrac{8}{18}}$

27. $\sqrt{\dfrac{(49)(16)}{25}}$

28. $\sqrt[3]{\dfrac{(27)(8)}{125}}$

For Exercises 29 through 34 solve the following equations for the unknown quantity (note: factoring can be used whenever it is appropriate). No rounding.

29. $x^2 = 4$

30. $y^3 = -27$

31. $p^6 = 64$

32. $m^4 = 16$

33. $w^{-2} = \dfrac{1}{25}$

34. $x^3 = 2$

1.3 BASICS OF SOLVING EQUATIONS

One major use of arithmetic operations on signed numbers and exponents is solving an equation for the unknown quantity. Any signed number that satisfies an equation (makes it true) is called a *solution* to the equation. For example, a value of x that satisfies the equation $-2x = 10$ is a solution to the equation. Similarly, a value of y that satisfies the equation $2 - y = 4$ is a solution to the equation.

An equation that has a solution is called a **conditional equation**, while one that does not have a solution is called an **inconsistent equation**. An equation, such as $x = x$, for which any number is a solution is called an **identity equation**.

In finding solutions (that is, one or more values that satisfy the equation), you should be aware of two notational conventions that are universally followed when writing equations with unknowns. First, parentheses are omitted for the product of a known number and an unknown quantity. For example, $(8)(y)$ is written as $8y$ and $(-3)(x)$ is written as $-3x$. Secondly, if the product involves a 1, the 1 is omitted and simply understood. Accordingly, both $(1)y$ and $1y$ are written as y, both $(1)x$ and $1x$ are written as x, both $(1)p$ and $1p$ are written as p, and so on. The same convention holds for -1. Thus, for example, both $(-1)y$ and $-1y$ are written as $-y$, and both $(-1)x$ and $-1x$ *are* written as $-x$.

A numerical value for an unknown quantity in an equation is a **solution** for the equation if that value, when substituted for the unknown, makes the equality valid. For example, to determine whether $x = 4$ is a solution of $-2x = 10$, substitute $x = 4$ into the equation. Because $-2x = -2(4) = -8$, which does not equal 10, the value 4 is not a solution to the equation.

Example 1 Determine whether or not $x = 2$ is a solution of the equation

$$\frac{5x + 3(x - 7)}{2x + 4} = -3.$$

Solution Substituting $x = 2$ into this equation, the left side becomes

$$\frac{(5)(2) + 3(2 - 7)}{2(2) + 4} = \frac{10 + 3(-5)}{4 + 4} = \frac{10 + (-15)}{8} = -\frac{5}{8}$$

Because this does *not* equal -3, which is the right side of the original equation, the proposed value of $x = 2$ is not a solution.

Example 2 Determine whether or not $p = 5/9$ is a solution of the equation $7 - p = 2 + 8p$.

Solution When $p = 5/9$ is substituted into this equation, the left side becomes

$$7 - \frac{5}{9} = \left(\frac{63}{9} - \frac{5}{9}\right) = \frac{58}{9}$$

and the right side becomes

$$2 + 8\left(\frac{5}{9}\right) = \frac{18}{9} + \frac{40}{9} = \frac{58}{9}$$

Because these values *are* equal, $p = 5/9$ is a solution.

One method for solving an equation for an unknown is trial and error. Guess a solution and then substitute it into the equation to see if it is valid. If not, guess again and continue guessing solutions until the correct one is found. Clearly, this method is time-consuming. It could take many guesses before the correct value is found, or one might never guess the correct value.

A more systematic procedure is to use algebra to isolate the unknown on one side of the equation. The correct value for this unknown will then be on the other side of the equation.

The Fundamental Rule of Algebra

To isolate an unknown on one side of an equation, you will be using the following fundamental rule:

Fundamental Rule of Algebra: *Whenever an arithmetic operation is performed on one side of an equation, an identical operation must be performed on the other side of the equation.* ***There are no exceptions to this rule****.*

This rule is frequently stated as the following three principles:

1. The *Addition/Subtraction* Principle:

For any equation $A = B$ and any real number n,

$$A + n = B + n$$

Notice that if n is a negative number, the same number is effectively subtracted from both sides of an equation.

2. The Multiplication/Division Principle:

For any equation $A = B$ and any real, non-zero number n,

$$(n)(A) = (n)(B)$$

and

$$\frac{A}{n} = \frac{B}{n}$$

It is important to emphasize that the number n either multiplies *every term* on *both sides* of the equation or divides *every term* on *both sides* of the equation, depending on which principle is used. For example, consider the equation $7x - 3 = 4x + 12$. Using the Multiplication Principle with $n = 2$ yields

$$2(7x - 3) = 2(4x + 12)$$

which becomes

$$14x - 6 = 8x + 24$$

3. **The Exponential Principle**:

For any equation $A = B$ and any real, non-zero exponent n,

$$A^n = B^n$$

Solving Equations Having One Unknown

The most useful applications of the fundamental rule of algebra are in finding solutions, if they exist, to equations having one unknown quantity

Example 3 Solve the equation $-2x = 10$ for x.

Solution The goal is to isolate x on one side of the equation. Here, this is accomplished by removing the multiplicative factor -2 in front of the x on the left side of the equation. This is easily accomplished using the multiplication/division principle and dividing both sides of the given equation by -2. Then,

$$\frac{-2x}{-2} = \frac{10}{-2}$$

which yields

$$x = -5.$$

Example 4 Solve the equation $x + 7 = 5$ for x.

Solution The goal is again to isolate x on one side of the equation. This is accomplished here by removing the additive factor $+7$ from the left side of the equation. By adding -7 to both sides of the equation (or, equivalently, by subtracting 7 from both sides of the equation, we obtain

$$x + 7 + (-7) = 5 + (-7)$$

which yields

$$x = -2.$$

Example 5 Solve the equation $2 - y = 4$ for y.

Solution We first remove the 2 from the left side, thereby leaving only y terms on that side of the equation. This is done by adding –2 to both sides of the equation (or, equivalently, by subtracting 2 from both sides of the equation). Then,

$$-2 + 2 - y = -2 + 4$$

which yields

$$-y = 2$$

We do not have y yet, but we are close. If we multiply both sides of the equation by –1, we obtain

$$(-1)(-y) = (-1)(2)$$

or

$$y = -2.$$

Example 6 Solve the equation $7 - p = 2 + 8p$ for p.

Solution We begin by grouping all the p terms on the same side of the equation. One way is to add p to both sides. Then,

$$7 - p + p = 2 + 8p + p$$
$$7 = 2 + 9p$$

Next we isolate the p terms on the right side by subtracting 2 from both sides of the equation

$$7 - 2 = 2 + 9p - 2$$

which yields

$$5 = 9p$$

Finally, we divide both sides of this equation by 9 to isolate p by itself. Thus, $5/9 = p$. Which can be rewritten as $p = 5/9$.

Example 7 Solve the equation $3(x - 7) = \dfrac{5x + 9}{4}$ for x.

Solution To eliminate the fraction, we multiply both sides of the equation by 4. Then,

$$4\left[3(x - 7)\right] = 4\left[\frac{5x + 9}{4}\right]$$

$$12(x - 7) = 5x + 9$$
$$12x - 84 = 5x + 9$$

Now, subtracting $5x$ from both sides, which eliminates the x term on the right side of the equation, yields

$$7x - 84 = 9$$

To isolate the x term, we now add 84 to each side of the equation, which yields

$$7x = 93$$

Finally, dividing both sides of the equation by 7 yields

$$x = 93/7$$

Example 8 Solve the following equation for x.

$$\frac{5x + 3(x - 7)}{2x + 4} = -3$$

Solution To eliminate the fraction, we multiply both sides of the equation by the quantity $(2x + 4)$, which yields

$$(2x + 4)\left[\frac{5x + 3(x - 7)}{2x + 4}\right] = (2x + 4)(-3)$$

which can be simplified to

$$5x + 3(x - 7) = -6x - 12$$

Then, multiplying the $(x - 7)$ term through by 3, we have

$$5x + 3x - 21 = -6x - 12$$

Adding the $5x$ to the $3x$ yields

$$8x - 21 = -6x - 12$$

Adding $6x$ to both sides yields

$$14x - 21 = -12$$

Adding 21 to both sides yields

$$14x = 9$$

Finally, dividing both sides by 14 yields the solution $x = 9/14$.

Exercises 1.3

For Exercises 1 through 6 determine whether or not the proposed values of the unknowns are solutions of the given systems.

1. $2x + 3 = 1;$ $x = -1$

2. $y + 4 = 2y;$ $y = 1$

3. $2(p + 7) = 3p + 4;$ $p = 1$

4. $x + 3 = 2(x + 1) + 1;$ $x = 0$

5. $\dfrac{(s+3)(s-2)}{2s+1} = s + 7;$ $S = 1.$

6. $\dfrac{(2t+3)(t-1)+1}{2(t+3)+1} = 3t - 4$ $t = 2;$

7. $3x - 7 = 5;$ $x = 7.$

8. $9y + 14 = 8y;$ $y = -14.$

9. $6(p+9) = 4 - 2(p-1);$ $p = -6.$

10. $\dfrac{(w+1)(w-2)}{2w-1} = w + 1;$ $w = 1.$

11. $\dfrac{(3m+1)(m-1)-12}{2(m-3)+1} = m + 3;$ $m = 8.$

12. Determine whether or not $x = 1$ is a solution of the following equation if it is known that $y = 2$ and $z = 0$:

$$\frac{x(y-1)+yz}{y(x-z)} = \frac{x}{y}$$

13. Determine whether or not $x = 2$ is a solution of the following equation if it is known that $y = -3$ and $z = 1.$

$$\frac{5(x+y)-xz}{y(z-x)} = \frac{3x+z}{y}$$

For Exercises 14 through 32, solve each equation for the unknown quantity.

14. $x + 7 = 2$

15. $7 = 2 + x$

16. $y - 8 = -2$

17. $8x = -16$

18. $2p = 30$

19. $-4p = 16$

20. $s + 10 = 2s$

21. $t - 10 = 4 - t$

22. $2t + 1 = t - 5$

23. $2x = 3(x + 1)$

24. $5y - 1 = 4(y - 2)$

25. $8(p - 2) = 7(2p + 1)$

26. $7p + 1 = 7(p - 1) + 3p$

27. $2(a + 7) - 4 = 3(a - 1) + 2a$

28. $\dfrac{(x-1)+5(x-4)}{8} = 2x + 1$

29. $\dfrac{2(y-1)+4}{y} = 8$

30. $\dfrac{(t-4)}{5} = \dfrac{3t+1}{8}$

31. $\dfrac{3(2t-6)+4(t-8)}{7(6+t)-8(t-4)} = -3$

32. $\dfrac{8t+9(1-t)+7}{3} = \dfrac{2(t-8)-5(t+1)}{2}$

1.4 SIGMA NOTATION

In various parts of this text, we will need the sum of large numbers of terms. Sometimes we will need the actual total, and then we will have to sum the numbers physically. Other times, however, we will only need to indicate the appropriate sum. An example is a statement, "Yearly expenditures are the sum of weekly expenditures." Here we are not explicitly calculating the total yearly expenditure but simply indicating that it is the sum of 52 numbers. In cases like this, there is a useful mathematical notation for indicating the appropriate sum.

Consider the case of a teacher who has a list of seven grades and wants their sum. If we denote the first grade as G_1, the second grade as G_2, and so on through the seventh grade, which we denote as G_7, the final sum can be given as

$$\text{Sum} = G_1 + G_2 + G_3 + G_4 + G_5 + G_6 + G_7 \qquad \text{(Eq. 1.7)}$$

Of course, if we had 52 items to add as opposed to only 7, writing an expression similar to Equation 1.7 would be tedious indeed. A more convenient way to represent the right side of Equation 1.7 symbolically as

$$\sum_{i=1}^{7} G_i$$

The capital Greek letter sigma (Σ) denotes a sum. The term(s) after the sigma, in this case, the letter G, tells us what values are to be added, in this case, we are adding G terms.

The quantity $i = 1$ at the bottom of the sigma indicates where the sum is to start from; in this case, the sum starts with G_1, that is, G_i, with i replaced with the starting value of 1. The number at the top of the sigma indicates where the sum is to stop, which in this case is G_7. Intermediate values in the sum are obtained by replacing the subscript i on individual G terms with consecutive integers between the starting and ending numbers at the bottom and top of the sigma, respectively.

For example, the notation

$$\sum_{i=3}^{8} S_i$$

indicates a sum of S terms. The sum starts with S_3, because the starting i value is given as 3 at the bottom of the sigma sign, and ends at S_8, because the ending value of i is given as 8 above the sigma sign. Thus, the notation

$$\sum_{i=3}^{8} S_i$$

is shorthand for

$$S_3 + S_4 + S_5 + S_6 + S_7 + S_8.$$

Additional examples are

$$\sum_{i=1}^{5} T_i = T_1 + T_2 + T_3 + T_4 + T_5$$

$$\sum_{i=2}^{4} x_i = x_2 + x_3 + x_4$$

and

$$\sum_{i=9}^{9} y_i = y_9$$

In general,

$$\sum_{i=m}^{n} q_i = q_m + q_{m+1} + q_{m+2} + \ldots + q_n, \quad m \le n \qquad \text{(Eq. 1.8)}$$

is the *sigma notation* for the sum of the quantities q_i ranging from q_m through q_n successively. Obviously, the left side of Equation 1.8 is more compact than the expanded form given on the right side, and is the reason for using sigma notation. When actual values are given for the terms being summed, a final value for the sum can be obtained.

Example 1 Give the expanded form of

$$\sum_{i=4}^{8} (x_i - i)$$

Solution The sigma notation indicated the sum of terms having the form $(x_i - i)$ beginning with $i = 4$ and continuing successively through $i = 8$. Therefore,

$$\sum_{i=4}^{8} (x_i - i) = (x_4 - 4) + (x_5 - 5) + (x_6 - 6) + (x_7 - 7) + (x_8 - 8)$$

Example 2 Give the expanded form of

$$\sum_{i=1}^{6} \left(\frac{i}{i+1} \right)$$

Solution

$$\sum_{i=1}^{6} \left[\frac{i}{(i+1)} \right] = \left(\frac{1}{1+1} \right) + \left(\frac{2}{2+1} \right) + \left(\frac{3}{3+1} \right) + \left(\frac{4}{4+1} \right) + \left(\frac{5}{5+1} \right) + \left(\frac{6}{6+1} \right).$$

Example 3 Determine the sigma notation for the sum

$$\left(\frac{3^2}{2}+1\right)+\left(\frac{4^2}{2}+1\right)+\left(\frac{5^2}{2}+1\right)+\left(\frac{6^2}{2}+1\right)+\ldots+\left(\frac{100^2}{2}+1\right)$$

Solution Each term in the sum is of the form $\left(\dfrac{i^2}{2}+1\right)$ where i is an integer starting at 3 and ending at 100, with all terms in between. The sum can be given as

$$\sum_{i=3}^{100}\left(\frac{i^2}{2}+1\right)$$

Example 4 Weekly expenditures for a given year are denoted as W_1 through W_{52} successively. Develop a formula for the yearly expenditure.

Solution Denote the yearly expenditure as Y. Because yearly expenditure is the sum of the individual weekly expenditures,

$$Y=\sum_{i=1}^{52}W_i$$

All the examples so far have used the subscript i. Any other letter would do equally well. Thus

$$\sum_{j=2}^{4}x_j=x_2+x_3+x_4$$

and

$$\sum_{k=1}^{10}k^2=1^2+2^2+3^2+4^2+5^2+6^2+7^2+8^2+9^2+10^2$$

Sometimes we are given data, and we would like to indicate that some portion of this data is to be summed. For example, if test scores for a particular student are 60, 70, 75, 80, 82, 83, 87, 90, and we only want to sum the 2nd through 7th scores for some reason, this is easily indicated by the notation $\sum_{i=2}^{7}G_i$. Of course, this notation is good only if we understand that G_1 signifies the first score, G_2 the second score, and G_7 the next to last score. In general, we always assume that data are ordered as they appear.

Finally, when we write a sigma without any numbers below or above it, we mean that the sum is to include all possible terms. It should be clear from the context which terms are being considered. For example, certain data may include pairs of numbers, and we may wish to multiply the members of each pair and then sum over all the pairs. If the number of data points is not known in advance, we can still indicate the desired sum using the notation $\Sigma x_i y_i$. For the data listed in Table 1.2, this sum is (because there are only four pairs of data points)

$$\Sigma x_i y_i = x_1 y_1 + x_2 y_2 + x_3 y_3 + x_4 y_4$$

$$= (5)(7) + (10)(12) + (14)(8) + (20)(15) = 567$$

TABLE 1.2

x	5	10	14	20
y	7	12	8	15

Exercises 1.4

1. Write the expanded form of the following expressions:

   a. $\displaystyle\sum_{i=1}^{3}(x_i)^2$

   b. $\displaystyle\sum_{i=3}^{11}2x_i$

   c. $\displaystyle\sum_{i=1}^{6}(x_i + y_i)$

   d. $\displaystyle\sum_{j=99}^{105}(3M_j + 4)$.

2. Write the expanded form of the following expressions:

   a. $\displaystyle\sum_{k=1}^{10}k$

   b. $\displaystyle\sum_{m=2}^{6}\left(\frac{m+2}{m+3}\right)$

   c. $\displaystyle\sum_{j=0}^{7}(j+1)$

   d. $\displaystyle\sum_{i=1}^{15}1$

   e. $\displaystyle\sum_{i=0}^{6}(-1)^i$

   f. $\displaystyle\sum_{p=7}^{14}(p-10)^2$

   g. $\displaystyle\sum_{i=1}^{8}i$

   h. $\displaystyle\sum_{k=2}^{6}\left(\frac{k}{k+1}\right)$

   i. $\displaystyle\sum_{j=0}^{7}(j-4)$

   j. $\displaystyle\sum_{m=1}^{5}\left[(-1)^m m\right]$

   k. $\displaystyle\sum_{p=1}^{10}(p-5)^2$

   l. $\displaystyle\sum_{m=1}^{5}3$

3. Write the following expressions in sigma notation.

   a. $3(2)^2 + 3(3)^2 + 3(4)^2 + 3(5)^2 + \ldots + 3(29)^2$

   b. $2(3)^2 + 3(3)^2 + 4(3)^2 + 5(3)^2 + \ldots + 29(3)^2$

c. $2(3)^2 + 2(3)^3 + 2(3)^4 + 2(3)^5 + \ldots + 2(3)^{29}$

d. $3(2)^2 - 3(3)^2 + 3(4)^2 - 3(5)^2 + \ldots + 3(28)^2 - 3(29)^2$

4. For the data given in Table 1.3, calculate the given sums.

TABLE 1.3

i	1	2	3	4	5
x	0	1	2	−1	4
y	6	7	8	3	−2

a. $\displaystyle\sum_{i=1}^{5} x_i$

b. $\displaystyle\sum_{j=1}^{5} y_j$

c. $\displaystyle\sum_{i=1}^{3} x_i$

d. $\displaystyle\sum_{k=1}^{5} (x_k + y_k)$

e. $\displaystyle\sum_{k=1}^{5}(x_k)\sum_{k=1}^{5}(y_k)$

f. $\displaystyle\sum_{m=2}^{4}(x_m y_m).$

g. What can you conclude about the sums found in parts d and e?

5. For the data given in Table 1.4, calculate the indicated sums

TABLE 1.4

i	1	2	3	4	5	6
x	0	8	−2	5	−3	7
y	3	2	6	9	10	1

a. $\displaystyle\sum x_i$

b. $\displaystyle\sum y_i$

c. $\displaystyle\sum (x_i)^2$

d. $\displaystyle\sum (x_i - 2)$

e. $\displaystyle\sum x_i y_i$

f. $\left(\displaystyle\sum x_i\right)\left(\displaystyle\sum y_i\right)$

g. What can you conclude about the sums found in parts e and f?

6. Write each of the following expressions in expanded form and verify that they are equal.

   a. $\displaystyle\sum_{i=1}^{6}\left(\frac{1}{i}\right)$

   b. $\displaystyle\sum_{i=2}^{7}\left(\frac{1}{i-1}\right)$

   c. $\displaystyle\sum_{i=0}^{5}\left(\frac{1}{i+1}\right)$

7. Prove the following identities by converting each side to expanded form:

   a. $\displaystyle c\left(\sum_{i=1}^{n}x_i\right)=\sum_{i=1}^{n}(cx_i)$

   b. $\displaystyle\sum_{i=1}^{n}(x_k+y_i)=\sum_{i=1}^{n}(x_i)+\sum_{i=1}^{n}(y_i)$

   c. $\displaystyle\sum_{i=1}^{m}x_i+\sum_{i=m+1}^{n}x_i=\sum_{i=1}^{n}x_i$

8. Determine if the following statement is valid or not.

$$\sum_{i=1}^{n}(x_iy_i)=\left(\sum_{i=1}^{n}x_i\right)\left(\sum_{i=1}^{n}y_i\right)$$

9. Derive a formula using sigma notation for the average of a set of grades $G_1, G_2, G_3, \ldots, G_n$

1.5 NUMERICAL CONSIDERATIONS

Expressing numbers in decimal form is necessary for most commercial transactions. One reason involves money. All financial figures are given in decimal form; the numbers to the left of the decimal point represent dollars, and the numbers to the right of the decimal point represent cents. Quoting the cost of an item as $1.20 is clearer than quoting six-fifths of a dollar. A second reason for using decimals is mathematical. It is easier to add 0.2 and 1.5 than it is to 1/5 and 1½. In this section, we present two considerations when dealing with numerical values; rounding when dealing with dollar and cents commercial transactions and exponential notation that you will sometimes encounter when using a calculator.

Rounding

The most common form of rounding is referred to as *arithmetic rounding* or just *rounding* for short. Here one first decides how many digits are to be retained and then looks at the next digit to the right. If it is greater than or equal to 5, the previous digit (which is the last one being kept) is increased by 1; otherwise, no change is made. All digits to the right are then either set to zero or dropped.

As an example, consider the irrational number $\pi = 3.1415926....$ To round this number to three decimal places, first look at the digit four places to the right of the decimal point. It is 5, so we increase the previous digit by 1 and write $\pi = 3.142$ rounded to three decimal places. To round the same number to two decimal places, first look at the digit three places to the right of the decimal point. It is 1, which is less than 5. Thus, $\pi = 3.14$ rounded to two decimal places.

Rounding is equally applicable to finite decimals. Anytime one uses the above procedure to reduce the number of digits in a number, one is rounding. For example, $81.314 = 81.31$ rounded to two decimal places, $8595.72 = 8596$ rounded to units, and $0.0051624 = 0.0052$ rounded to four decimal places.

The second form of rounding is called *rounding up*. Here the last digit being kept in a number is automatically increased by 1 if *any one* of the discarded digits is not 0. For example, $8.1403 = 8.15$ rounded up to two decimal places (note that one of the discarded digits is a 3; that is, not all of the discarded digits are zero), $1.38112 = 1.382$ rounded up to three decimal places, and $1.900 = 1.9$ rounded to one decimal place. In the last example, we did not increase the 9 because all the discarded digits were 0.

Rounding up is used by banks for mortgage payments and merchants for determining prices. If a mortgage payment is $241.5723 exactly, a bank will charge $241.58. If the retail price of a product is $1.2905, a merchant will typically charge $1.30. Rounding up is employed when the user does not wish to absorb the losses incurred by regular arithmetic rounding.

Still, the third form of rounding is *rounding down*, which is more commonly known as *truncation*. Here the last digit being kept is never changed, regardless of the magnitude of the numbers being discarded. Thus, $89.318 = 89.31$ truncated to two decimals, and $13.75 = 13$ truncated to units.

Truncation is often used in reporting the number of finished goods produced. At the end of a day, a company may have produced $82\frac{5}{8}$ cars, but reports production of only 82 completed cars.

Exponential Notation

Calculators that have the ability to display more decimal values than can be accommodated on their displays typically display both very small and very large numbers using exponential notation. In this notation, the letter E stands for "exponent" and the number following the E indicates the number of places the decimal point should be moved to obtain the standard decimal value. The decimal point is moved to the right if the number after E is positive, or it is moved to the left if the number after the E is negative

For example, the E10 in a display such as 1.625 E10 means move the decimal place ten places to the right, so the number becomes 16250000000. The E-8 in a display such as 7.31 E-8 means move the decimal point eight places to the left, so the number becomes .0000000731. Table 1.5 provides a number of additional examples using exponential notation.

TABLE 1.5

Exponential Notation	Decimal Notation
2.689 E5	268900.
2.689 E-5	.00002689
4.896723 E10	48967230000.
4.896723 E-10	.0000000004896723

Exercises 1.5

In Exercises 1 through 5, use a calculator to find the equivalent decimal values for the given fraction, and then round (that is, use arithmetic rounding) the numbers to two decimal places.

1. 2/3. **2.** 4/11 **3.** 4/17 **4.** 12/7 **5.** 89/31

6. Repeat Exercises 1 through 5, but round the numbers to three decimal places.

7. Round up the numbers given in Exercises 1 through 5 to three decimals.

8. Truncate the numbers given in Exercises 1 through 5 to two decimal places.

In Exercises 9 through 14, convert the numbers written in exponential notation to standard decimal numbers:

9. 3.35642 E3 **10.** 5.62384 E9 **11.** 3.356 E-3

12. 8.63 E-7 **13.** 3.3 E12 **14.** 4.3 E-12

1.6 SUMMARY OF KEY POINTS

Key Terms

- Exponential notation
- Exponents
- Fundamental Rule of Algebra
- Order of operations
- Sigma Notation
- Signed numbers
- Solution

Key Concepts

- Numbers can be positive, negative, or zero.
- The product of two positive numbers is positive.
- The product of two negative numbers is positive.
- The product of two different signed numbers is negative.
- The order of operations is the set of rules defining how any mathematical expression consisting of more than two numbers must be calculated. This order is as follows:
  - All values within parentheses must be evaluated first, from the innermost to the outermost sets of parentheses.
  - All numbers raised to a power (exponents) are calculated.
  - All multiplications and divisions are calculated, from left to right.
  - All additions and subtractions calculated, from left to right.
- Exponents are a convenient notation for representing the product of a number times itself.
- The Fundamental Rule of Algebra is that any mathematical operation performed on one side of an equation must also be performed on the other side of the equation.
- A solution to an equation is any signed number that makes the equation true.
- An equation having one unknown can be solved by repeatedly applying the fundamental rule of algebra.
- Sigma notation is used to indicate the sum of similar mathematical terms.
- Exponential notation is used to indicate either very small or very large numbers concisely.

GRAPHS AND LINEAR EQUATIONS

In this Chapter

Graphs have a visual impact and are a powerful medium for presenting data, and relationships, just as pictures do for words. They are used by the Census Bureau to display national statistics, by newspapers to show trends, and by social scientists to demonstrate correlations between social behaviors. Corporations use graphs to illustrate growth, and speakers routinely use graphs to support their conclusions. Graphs cut across all disciplines as the most widely used model for displaying real-world behavior.

Equations are a convenient and concise way of representing relationships between quantities, such as sales and advertising, profit and time, cost and number of units manufactured, and so on. As such, they are used to "model" or represent real-world situations. Although an equation usually does not reveal all the relationships between the quantities under investigation, they often contain enough information for one to make meaningful observations and practical decisions.

2.1 THE CARTESIAN COORDINATE SYSTEM

A graph is a diagram that represents data in an organized manner. Bar graphs, column graphs, line graphs, and pie charts, as shown in Figures 2.1 through 2.3, respectively, illustrate these types of graphs.[1]

In this chapter, we will focus our attention on line graphs, although, as shown in Section 2.6, this type of graph is easily converted into any of the other graph types shown in Figures 2.2 and 2.3.

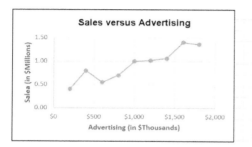

FIGURE 2.1 Two examples of line graphs.

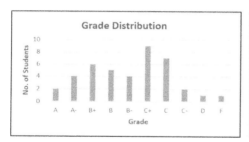
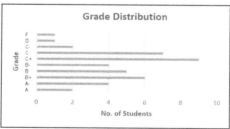

FIGURE 2.2 (a) A column graph.   FIGURE 2.2 (b) A bar graph.

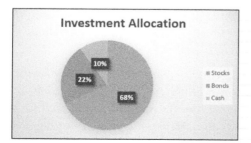
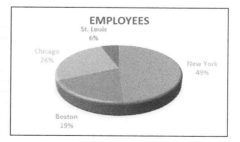

FIGURE 2.3 (a) A 2-dimensional pie chart.   FIGURE 2.3 (b) A 3-dimensional pie chart.

---

[1] How these graph types are constructed using Excel® is presented in Section 2.6.

The most commonly used line graph is constructed as points on a two-dimensional plane constructed on a Cartesian coordinate system. To understand this system, first, consider the map illustrated in Figure 2.4 with streets running in either a north–south direction or an east–west direction.

By using the center of Broad and Market Streets as a reference (perhaps a motorist has stopped there for directions), it is easy to locate any other point on the map. The intersection of Elm Lane and Maple Street is two blocks west and one block south of the reference point. The light at Freeman Street and Valley Road is three blocks east and two blocks north of the reference point.

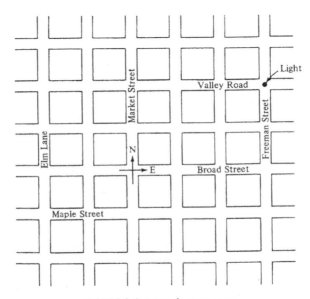

FIGURE 2.4 A sample street map.

In each case, the new point on the map is uniquely determined from the reference point by two numbers and their direction.

The *Cartesian coordinate system*, which is also known as the *Rectangular coordinate system*, is a generalized version of the previous map. To construct this system, two intersecting perpendicular lines (forming an angle of 90 degrees with each other) are first drawn, as illustrated in Figure 2.5. The horizontal line is often called the x-coordinate axis (or just the x-axis for short), while the vertical line is often called the y-coordinate axis (or y-axis for short). The intersection of these two axes is the *origin*, and it represents the reference point of the system.

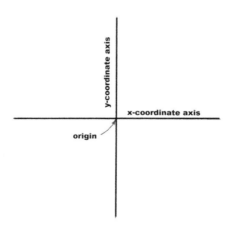

FIGURE 2.5 Perpendicular intersecting lines.

Each axis is marked in fixed units of length, as illustrated in Figure 2.6 Units to the right of the origin on the axis and units above the origin on the *y*-axis are assigned positive values. Units to the left of the origin on the *x*-axis and units below the origin on the *y*-axis are assigned negative values. The complete system is the *Cartesian* (or rectangular) *coordinate* system.

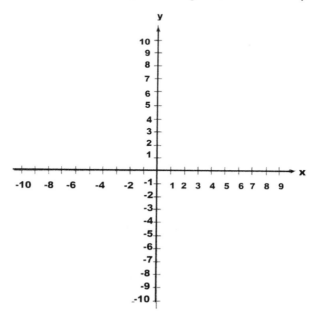

FIGURE 2.6 A sample Cartesian coordinate system.

Note that arrows have been appended to the positive portions of the x- and y-axes in Figure 2.6. These arrows simply indicate, visually, the direction of increasing values of x and y. We can think of the x-axis as representing the east–west direction on a map, and the y-axis as representing the north–south direction. The arrows then indicate the directions east and north. Moving in a positive x-direction corresponds to moving east. Moving in a negative y-direction corresponds to moving south. Motion in the directions of north and west is defined by moving in the positive y and negative x directions, respectively.

The usefulness of the Cartesian coordinate system is that any point on the plane can be located from the origin by two numbers. Directions need not be specified since they are inherent in the signs of the numbers. As an example, consider point P shown in Figure 2.7. To reach this point from the origin, one must move 6 units along the x-axis in the positive direction, and then 4 units in the positive y-direction along a line segment parallel to the y-axis *beginning* at $x = 6$. The point P is located by the two numbers 6 and 4, if we agree that the first number, 6 denotes moving 6 units along the x-axis and the second number, 4, denotes moving 4 units in the y-direction.

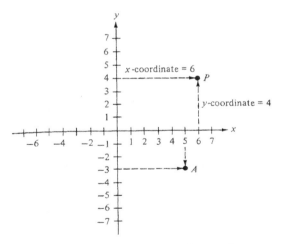

FIGURE 2.7 Locating a point on a Cartesian coordinate system.

Every point on the plane is uniquely determined by an ordered pair of two numbers in the form (x, y), which together are called the *coordinates* of the point.

The first number in an ordered pair is called the *x-coordinate*, and the second number, the *y-coordinate*. The coordinates of point P in Figure 2.7 are

(6, 4); its x-coordinate is 6 and its y-coordinate is 4. The coordinates of point A in the same figure are (5, −3); its x-coordinate is 5 and its y-coordinate is −3.

Example 1 Using a Cartesian coordinate system, locate and plot the point having coordinates (3, −5).

Solution The point of interest is reached by first moving 3 units along the x-axis in the positive direction from the origin and then moving 5 units in the negative y-direction. The point is plotted as Q in Figure 2.8.

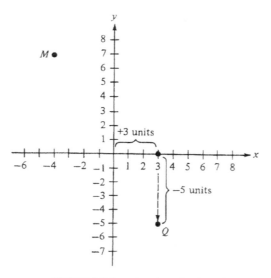

FIGURE 2.8 Locating the point (3, −5).

Example 2 Determine the coordinates of the point M shown in Figure 2.8.

Solution To reach M from the origin, we must move 4 units in the negative x-direction (which implies an x-coordinate of −4) and then 7 units in the positive y-direction (which implies a y-coordinate of +7). The coordinates are (−4, 7). Recall that the x-coordinate is always given before the y-coordinate.

A Cartesian coordinate system divides the plane into four sections. Each section is called a *quadrant*, labeled I, II, III, or IV, as illustrated in Figure 2.9. A point in the first quadrant has both a positive x- and a positive y-coordinate. A point in the second quadrant has a negative x-coordinate and a positive y-coordinate, while a point in the third quadrant has a negative x-coordinate and a negative y-coordinate. A point in the fourth quadrant has a positive x-coordinate but a negative y-coordinate.

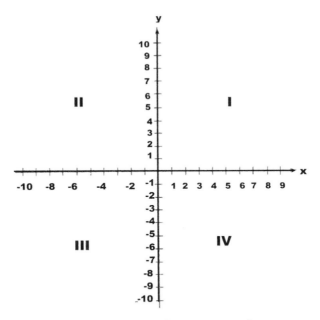

FIGURE 2.9 The four quadrants of a Cartesian coordinate system.

In practice, two modifications are often made in the Cartesian coordinate system as we have defined it. First, the letters used to label the axes need not be *x* and *y* but can be any other two letters that are convenient. In a problem dealing with the price and demand of a certain product, it may be more revealing to label the axes *P* for price and *D* for demand. Regardless of the letters used, the first component of a point always refers to a movement along the horizontal axis (i.e., the conventional *x*-axis), while the second component refers to movement in a direction parallel to the vertical axis (i.e., the conventional *y*-axis).

The second modification is that the same scale need not be used on both coordinate scales. The scale on the horizontal axis can, and frequently does, differ from the scale used on the vertical axis. Figure 2.10 illustrates this, where each axis uses a different scale. The coordinates of point B on this figure are (2, 25). *The only restriction in this modification is that each scale, individually, once chosen, must be marked off consistently in equal units.*

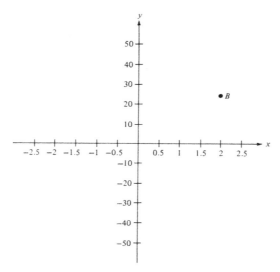

FIGURE 2.10 The *x*- and *y*-axes can have differing scales.

Exercises 2.1

1. Consider the points labeled *A* through *K* in Figure 2.11

   a. Determine the coordinates of each point.

   b. Which points are located in quadrant I?

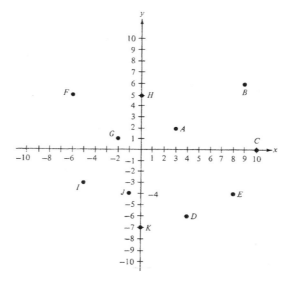

FIGURE 2.11 Through 2.13 do not need captions, as they are used in the Exercises.

2. Redo Exercise 1 for the points *A* through *H* in Figure 2.12.

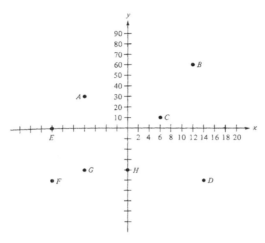

FIGURE 2.12

3. Consider the points shown in Figure 2.13

    a. Draw a coordinate system that places *A* in quadrant I, *B* in quadrant II, and *C* in quadrant IV.

    b. Draw a coordinate system that places *A*, *B*, and *C* in quadrant I and *D* in quadrant IV.

    c. Draw a coordinate system that places all points in quadrant III.

FIGURE 2.13

4. Plot each of the following points on the same coordinate system:

    a. $(3, -1)$ **b.** $(2, 5)$ **c.** $(-5, 2)$ **d.** $(-6, -6)$

5. Plot each of the following points on the same coordinate system:

    a. $(3, -10)$ **b.** $(2, 50)$ **c.** $(-5, 20)$ **d.** $(-6, -60)$

6. Plot each of the following points on the same coordinate system:

   a. $(250, 45)$ **b.** $(500, -10)$ **c.** $(400, 20)$ **d.** $(-6, -60)$

7. Plot each of the following points on the same coordinate system:

   a. $(2, 0)$ **b.** $(3, 5)$ **c.** $(-4, -7)$

8. Plot each of the following points on the same coordinate system:

   a. $(-6, 9)$ **b.** $(0, -10)$ **c.** $(8, -3)$

9. **a.** Determine the value of the y-coordinate of every point on the x-axis.

   b. Determine the value of the x-coordinate of every point on the y-axis.

10. Construct a Cartesian coordinate system with each axis scaled the same. Draw a straight–line segment between the origin and the point having coordinates $(10, 10)$. Determine the angle this line makes with the positive x-axis.

11. Construct a Cartesian coordinate system and draw a line parallel to the x-axis. What do all points on this line have in common?

2.2 LINE GRAPHS

Line Graphs are most often displayed on a *Cartesian coordinate system,* which was presented in the previous section. As previously described, this coordinate system consists of two intersecting perpendicular lines called *axes*, as was shown in Figure 2.5 and reproduced below as Figure 2.14 for convenience.

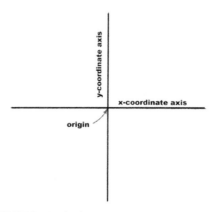

FIGURE 2.14 A basic Cartesian coordinate system.

Recall from Section 2.1 that he intersection of the two axes is the Origin, which is the reference point for the system. The horizontal line is typically called the x-axis and the vertical line is typically called the y-axis. Although the letters x and y are the most widely used symbols for these axes, other letters are used when they are more meaningful in a particular problem.

Tick marks are then used to divide each axis into fixed units of length. Units to the right of the origin on the x-axis and units above the origin on the y-axis are assigned positive values. Units to the left of the origin on the x-axis and units below the origin on the y-axis are assigned negative values. Arrowheads are affixed to the positive portions of the x- and y-axes to indicate the directions of increasing values of x and y. Successive tick marks must be equally spaced, which makes the units between successive tick marks the same, although the units or *scale* on the horizontal axis can, and frequently does, differ from the scale used on the vertical axis.

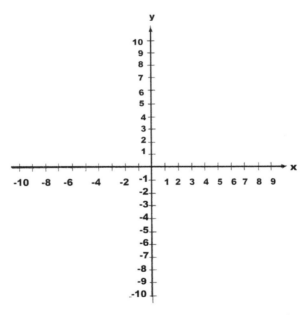

FIGURE 2.15 A complete Cartesian coordinate system with tick marks and arrowheads.

Every point in the plane defined by a Cartesian coordinate system is uniquely determined by an ordered pair of numbers, such as (2,5) and (−3,7), which are called the *coordinates* of the point. The first number in the parenthesis the *x-coordinate* and the second the *y-coordinate*. Thus, the point defined by the coordinates (2,5) has an x-coordinate of 2 and a y-coordinate of 5.

Example 1 Graph the data in Table 2.1, which lists the amount of rainfall versus the yield of a particular plant.

TABLE 2.1 Rain versus Yield.

Rain (Inches)	1	1.7	2.3	3.2	3.8	4.3	4.8
Yield (Bushels)	9	18	25	30	34	40	45

Solution Table 2.1 provides the following seven coordinate points:

$$(1, 9), (1.7, 18), (2.3, 25), (3.2, 30), (3.8, 34), (4.3, 40), (4.8, 45)$$

Figure 2.16 illustrates the graph produced by these points. (The step-by-step procedure for creating this graph using Excel is presented in Section 2.6.)

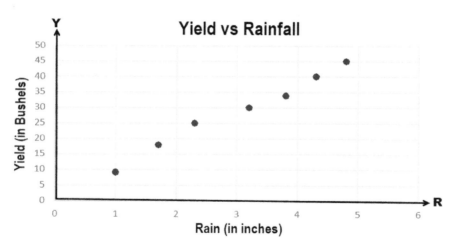

FIGURE 2.16 A line graph for yield versus rainfall.

Equations and Graphs

Graphs can be extremely useful in showing the solutions to an equation having two unknowns, such as the equations $y = 3x - 2$ or $y = 2x^2 + 3$. Any ordered pair of two numbers (x, y) that satisfy the equation (makes it true) is called a *solution* to the equation. Here the first value in parenthesis refers to the x value and the second value to the y variable.

Example 2 Determine if the ordered pair $(3, 7)$ is a solution to the equation

$$y = 3x - 2.$$

Solution Substituting the value $x = 3$ and $y = 7$ into the equation yields

$$7 = 3(3) - 2 = 7$$

As this is a true statement, the ordered pair $(3, 7)$ is a solution to the given equation.

The difficulty with equations in two unknowns is generally they have not one, but infinitely many solutions. Some of the solutions can be found by arbitrarily selecting values for one of the unknowns and substituting these values into the given equation, and then solving for the remaining unknown.

As an example, again consider the equation $y = 3x - 2$. Arbitrarily setting $x = 1$ and substituting it into the equation, we find $y = 3(1) - 2$ or $y = 1$. Therefore $x = 1$ and $y = 1$, represented as the point $(1, 1)$, is a solution to the equation.

To obtain another solution, a different value for x can arbitrarily be selected, and then the resulting equation solved for y. For example, setting $x = 2$ and substituting it into the given equation, yields $y = 3(2) - 2$ or $y = 4$. Therefore, $x = 2$ and $y = 4$, represented as the point $(2, 4)$ is a second solution. Table 2.2 lists a number of these solutions, where each y value is listed immediately below its corresponding x value.

TABLE 2.2 Solutions to the Equation $y = 3x - 2$.

x	−2	−1	0	1	2	3
y	−8	−5	−2	1	4	7

The main objection to this procedure is that it provides only *some* of the solutions, not all of them. Yet, it is not likely that we can ever do better algebraically because one equation having two unknowns generally has infinitely many solutions. If we are willing to use graphical methods, however, the prospects are brighter.

Every solution of one equation in two unknowns is a pair of numbers, such as $x = 2$ and $y = 4$, or $x = 3$ and $y = 7$ listed in Table 1.1 for the equation $y = 3x - 2$. Each solution pair then can be plotted as a single point on a Cartesian coordinate system. For example, Figure 2.17 shows the resulting plot of the five solution pairs listed in Table 2.1.

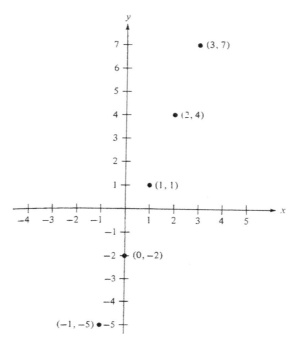

FIGURE 2.17 Plotting the points listed in Table 2.2.

Graphically, the set of *all* solutions to a single equation in two unknowns becomes a curve. Typically, the exact shape of the curve often can be determined by looking at a plot of some of the solution points and then making an educated guess. In particular, the points shown in Figure 2.17 appear to lie on a straight line; it seems likely, therefore, that the geometric solution of $y = 3x - 2$ is the dashed line drawn in Figure 2.18. We have drawn the line dashed to indicate that, at this stage, we cannot be certain that every point on the dashed straight line has coordinates that are solutions to the given equation. For example, the curve illustrated in Figure 2.19 also contains the points listed in Table 2.2.

One way to be certain that Figure 2.18 is correct is to plot more points than the five listed in Table 2.2. (However, by using methods presented in the next chapter, it can easily be verified that Figure 12.18 is, in fact, the correct graph of the equation $y = 3x - 2$ without plotting additional points.) For now, however, we must plot as many points as necessary to gain a reasonable idea of the shape of the curve before drawing it.

Let us summarize our steps. Given an equation with two unknowns, in this case, the equation $y = 3x - 2$, we want to find the set of all solutions. As a first step, some solutions are found by arbitrarily picking values for one of the unknowns, substituting these values into the given equation, and solving the resulting equation for the corresponding values of the other unknown.

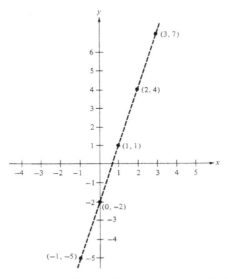

FIGURE 2.18   One possible curve for the points plotted in Figure 2.17.

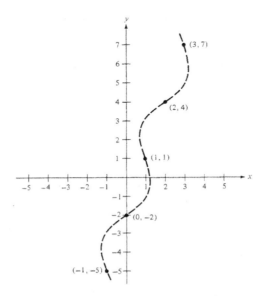

FIGURE 2.19 A second curve for the points plotted in Figure 2.17.

The second step was to plot every solution that was found, and the last step was to determine the curve that contained the plotted points. At this stage, a dashed line is constructed *that appears* to fit the plotted points best. Thus, the line plotted in Figure 2.17 is the graph of the equation $y = 3x - 2$.

Example 3 Find the graph of the equation $y = 2x^2 + 1$.

Solution We find solutions to this equation by arbitrarily assigning values to x and then finding the corresponding values for y Thus,

$$\text{when } x = 0, y = 2(0)^2 + 1 = 1$$

and

$$\text{when } x = -1, y = 2(-1)^2 + 1 = 3$$

Continuing in this manner, we generate Table 2.3

TABLE 2.3 Solutions to the Equation $y = 2x^2 + 1$.

x	−3	−2	−1	0	1	2	3
y	19	9	3	1	3	9	19

Plotting these points on a Cartesian coordinate system and then connecting the points. with a curve, we obtain Figure 2.20, which is the graph of $y = 2x^2 + 1$.

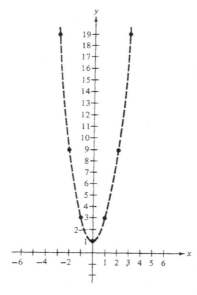

FIGURE 2.20 The graph of the equation $y = 2x^2 + 1$.

When solving an equation having two unknowns, you get to decide which variable to assign value to and which variable to calculate. For example, in the equation $y = 2x^2 + 1$, it is easier to select an x value and then solve for the y value rather than the reverse. However, for an equation such as $y^5 - 2y^2 = x + 1$, it is a simple matter to pick value of y and then solve for x, but much more difficult to solve for y-values if x-values are first selected

A few observations are now in order. First, the graph of an equation is only a geometric representation of the solutions of one equation in two unknowns. The actual solutions are the coordinates of each and every point on the graph. Thus, Figure 2.20 is the geometric representation of the solutions to $y = 2x^2 + 1$. The solutions themselves are the coordinates of each and every point on the curve.

Secondly, the curves plotted in Examples 2 and 3 were determined by looking at only a few points and then assuming that the curve behaved nicely between the plotted points. Formal justification of this assumption, however, requires the material considered in the next section. For now, we will continue to assume that a curve can be drawn when a sufficient number of points has been plotted. Again, just how many points is sufficient depends on the shape of the curve and the foresight of the plotter, and as indicated previously, "too many is always better than too few."

Example 4 Graph the equation $y = x^3 - x^2 + 1$

Solution For this equation it is easier to select value for x and then solve for y, rather than the reverse. In particular,

$$\text{when } x = 0, y = 1$$

and when $x = -2$, $y = (-2)^3 - (-2)^2 + 1 = -8 - 4 + 1 = -11$

Continuing in this manner, we generate Table 2.4

TABLE 2.4 Solutions to the Equation $y = x^3 - x^2 + 1$.

x	−2	−1	0	.5	1	2
y	−1 −1	3	1	.875	1	5

Plotting these points on a Cartesian coordinate system and then connecting the points with a curve, we obtain Figure 2.21.

Note: Graphing calculators and software packages such as Excel provide tools for evaluating equations and graphing the solutions. Use of these tools, when they are available, is highly recommended. Section 2.6 presents the steps necessary to produce graphs using Excel.

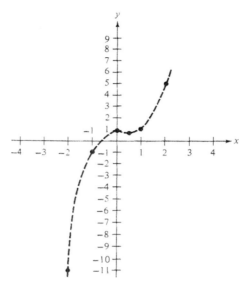

FIGURE 2.21 The graph of the equation $y = 2x^2 + 1$.

Example 5 Graph the equation $x^2 + y^2 = 25$.

Solution We can either select values for y and solve for x, or select values for x and solve the equation for y. As both paths are equal in difficulty, we arbitrarily select y first and rewrite the equation as

$$x = \pm\sqrt{25 - y^2}$$

when $y = -4$,

$$x = \pm\sqrt{25 - y^2} = \pm\sqrt{25 - (-4)^2} = \pm\sqrt{25 - 16} = \pm 3$$

Continuing in the manner, we generate Table 2.5

TABLE 2.5 Solutions to the Equation $x^2 + y^2 = 25$.

y	−5	−4	−3	0	3	4	5
x	0	±3	±4	±5	±4	±3	0

Plotting these points, we obtain Figure 2.22 as the graph of the given equation.

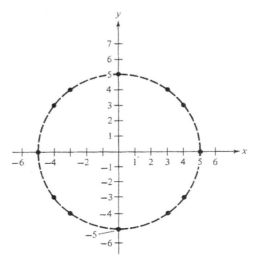

FIGURE 2.22 The graph of the equation $x^2 + y^2 = 25$.

Piecewise Graphs

Although graphs can plot observed points (see Figure 2.16) and others visually present solutions to a single equation (see Figures 2.20 to 2.21), graphs are also used for situations described by two or more equations. Figures 2.23 is an example of one such graph.

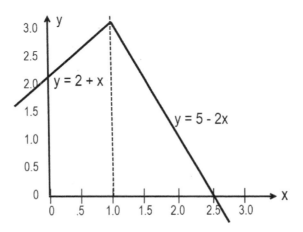

FIGURE 2.23 A graph consisting of two distinct lines.

Figure 2.23 is an example of a class of graphs known as *piecewise graphs*. A piecewise graph is defined by two or more equations, specified on non-overlapping intervals. The piecewise graph shown in Figure 2.22 is defined by the two equations

$$y = \begin{cases} 2+x & x < 1 \\ 5-2x & x \geq 1 \end{cases}$$

Piecewise graphs neither require that individual pieces of the graph are straight lines nor that individual lines touch each other. For example, Figure 2.24 shows a piecewise graph consisting of two non-touching straight lines.

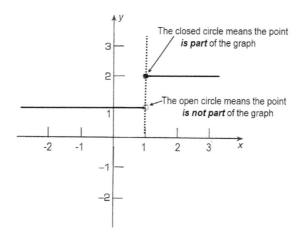

FIGURE 2.24 A piecewise graph with a "gap" or ledge.

As shown in Figure 2.24, a closed circle indicates a point that is included in the graph and an open circle indicate it is not. This notation is required to visually specify the y value associated with a common x value. This value would also be clearly defined by the equations defining the graph. For example, the defining equations for Figure 2.24, which indicate the y value to be used when x is 1 are:

$$y = \begin{cases} 1 & x < 1 \\ 2 & x \geq 1 \end{cases}$$

Example 5 A manufacturer provides a discount for large order purchases. Construct a graph for the discount structure listed in Table 2.6.

TABLE 2.6 Discount Table.

Quantity Ordered	Discount
Upto 100	2%
100 upto 200	4%
200 upto 300	6%
300 and above	8%

Solution Figure 2.25 is a graph of the discount structure listed in Table 2.6.

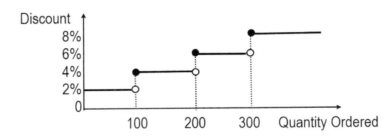

FIGURE 2.25 A discount versus quantity piecewise graph.

Exercises 2.2

Plot the graphs of the equations in Exercises 1 through 10.

1. $2x - 3y = 5$.   **2.** $y = x - 2$.

3. $6x - 2y = 3$.   **4.** $x^2 - y = 0$.

5. $y = 2x^2$.   **6.** $3y - 4x = 7$.

7. $y = x^3 - 2x^2 + x$.   **8.** $y = 3x - 4$.

9. $x^2 - y^2 = 0$.   **10.** $y = \sqrt{x}$

11. Graph $y = x$ and $y = -x$ on the same axes. How do these curves differ?

12. Graph the equation $y = 2x + 5$. Select x-values of 0, 1, 3, 5, 7 and $-1, -3, -5, -7$. Determine from the graph the values of y when x is 2 and when x is -8.

13. Determine if the points having coordinates $(-2, 6)$, $(0, 2)$, and $(1, 9)$ lie on the graph of the equation $y = 3x^2 + 4x + 2$.

14. Graph the two equations $y = 6x + 3$ and $y = 5x - 2$ on the same coordinate system. Determine the point of intersection of these curves from the resulting graph.

15. Graph the piecewise curve defined by the following equations:

$$y = \begin{cases} 2x + 4 & x < 1 \\ 1 - x & x \geq 1 \end{cases}$$

16. Graph the piecewise curve defined by the following equations:

$$y = \begin{cases} x & x \leq 2 \\ -x & x > 2 \end{cases}$$

17. Graph the piecewise curve defined by the following equations:

$$y = \begin{cases} x & x \leq 2 \\ 3 & x > 2 \end{cases}$$

18. Graph the piecewise curve defined by the following equations:

$$y = \begin{cases} 2x + 2 & x \leq 1 \\ 4x + 1 & x > 1 \end{cases}$$

19. Graph the piecewise curve defined by the following equations:

$$y = \begin{cases} 2 & x < 1 \\ 4 & 1 \leq x < 3 \\ 6 & x \geq 3 \end{cases}$$

2.3   GRAPHING LINEAR EQUATIONS

Graphing an equation can be considerably shortened if we already know the shape of the curve that the equation describes. One such case is provided by *linear equations*. These equations are singled out because they are one of the simplest and most important equations in both business and mathematics. They also have the geometric property that their graphs are always straight lines, among other useful features that are presented in the next section.

Formally, an equation is *linear* in two variables, x and y, if it satisfies the following definition:

Definition 2.1: A linear equation in two variables x and y is an equation of the form:

$$Ax + By = C \qquad \text{(Eq. 2.1)}$$

where A, B, and C are known real numbers and A and B are not both zero (this avoids the equations of the form $0 = C$. The variables x and y can be replaced by any other convenient letters. Thus, if x and y are replaced by p and q, respectively, then Equation 2.1 becomes the linear equation $Ap + Bq = C$ in the variables p and q.

Definitions are very precise mathematical statements. Unfortunately, this precision often makes a definition seem very complicated when, in fact, it is not. Usually, a few moments of thought is all that is needed to convert the given statement to an understandable concept.

For example, Definition 2.1 simply states that any equation having the form of (i.e., looks like) the equation $Ax + By = C$ where the letters A, B, and C are replaced by numbers (e.g., $3x + 7y = 10$), is called a *linear* equation. Another important point is that because an exponent of 1 is understood but not written, a necessary feature of a linear equation is that the exponent of both the x and y terms must only be 1.

The definition does not give any clues as to what a linear equation means geometrically-that will come later. What it does say, however, is that any equation that can be written in a form that looks like Equation 2.1, is a linear equation.

Example 1: Show that each of the following equations is linear.

a. $-x + 7y = 0$ **b.** $7y = 0$

c. $2x + 4 = 3y$ **d.** $\dfrac{1}{2}N - \dfrac{3}{4}P = 1.7$

Solution

a. This equation has the form $Ax + By = C$, with $A = -1$, $B = 7$, and $C = 0$. As such, it is a linear equation.

b. This equation has the form $Ax + By = C$, with $A = C = 0$ and $B = 7$. As such, it is a linear equation.

c. This equation can be rewritten as $2x - 3y = -4$, which has the form of Equation 2.2 with $A = 2$, $B = -3$, and $C = -4$. As such, it is a linear equation.

d. This is a linear equation in the variables N and P. As t has the form of Equation 2.1 with $A = 1/2$, $B = -3/4$, and $C = 1.7$, it is a linear equation.

Example 2 Determine whether or not $x^2 + y^2 = 4$ is a linear equation.

Solution For an equation to be linear, it must either have the form, or be able to be put into the form, of Equation 2.1. This equation cannot be put into that form, because both quantities x and y are squared, whereas Equation 2.1 requires both x and y to appear by themselves, with an implicit exponent of one and multiplied only by known numbers.

Example 3 Determine whether or not $\frac{1}{x} + 2 = 0$ is a linear equation.

Solution No, this is not a linear equation. Here x appears as $\frac{1}{x}$ and not as x multiplied by a known number as required in Equation 2.1.

Equation 2.1 is indeed very precise. An equation is called a linear equation *if and only if* it has the form of a constant times one quantity raised to the 1st power plus a constant times another quantity raised to the 1st power, with the sum of both terms equal to a constant. No x^2 terms, no $\sqrt{x}$ terms, and no xy terms are allowed.

It should also be stressed that the letters x and y in Equation 2.1 are *not* fixed; any two letters can be used. Thus $6C + 2N = 15$ is a linear equation in the variables C and N, and $0.5p - 0.75q = 1.7$ is linear equation in the 'quantities p and q. Linear equations are singled out as a special class of equations because they have several useful properties. One of these properties is that the graphs of all linear equations are straight lines (we will examine the remaining properties in Section 2.3. To illustrate this straight line property, a number of linear equations will first be graphed by plotting multiple points, as was done in the previous section.

Example 4 Graph the equation $6x + 2y = 15$.

Solution We initially plot some points satisfying the equation by arbitrarily selecting values of either x or y and finding the corresponding values of the other variable in the equation. Here we will select x-values and solve for the corresponding y-values. When $x = 0$, Equation 2.2 becomes $6(0) + 2y = 15$, resulting in $y = 7.5$. When $x = 1$, Equation 2.2 becomes $6(1) + 2y = 15$ and, solving for y, we obtain $y = 4.5$. Continuing in this manner, we generate Table 2.7.

TABLE 2.7 Solutions to the Equation $6x + 2y = 15$.

x	−1	0	1	2	3
y	10.5	7.5	4.5	1.5	−1.5

Plotting these points and connecting them produces the graph shown in Figure 2.26.

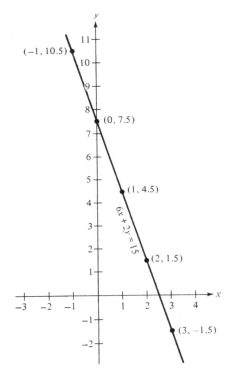

FIGURE 2.26 The graph of the equation $6x + 2y = 15$.

Example 5 Graph the equation $-x + 7y = 0$.

Solution Again we first plot points on the curve by selecting arbitrary values for either x or y and then finding their corresponding y or x-values. Arbitrarily selecting x-values, we compute as follows. When $x = 0$, the equation becomes $-(0) + 7y = 0$, hence $y = 0$. When $x = 1$, the equation becomes $-(1) + 7y = 0$ from which we obtain $y = 1/7 = 0.142857$, which is 0.15 rounded to two decimal places. Continuing, we generate Table 2.8 and Figure 2.27

TABLE 2.8 Solutions to the Equation $x + 7y = 0$.

x	0	1	2	3	7
y	0	0.15	0.29	0.43	1

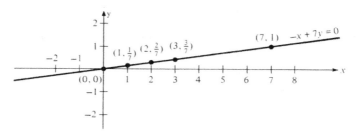

FIGURE 2.27 The graph of the equation $x + 7y = 0$.

In Examples 4 and 5 we arbitrarily selected x-values and then solved for the corresponding y-values. The same graphs would be produced if we had first picked y-values and then solved for the corresponding x-values.

Example 6 Redo Example 4 by selecting y-values and solving for the corresponding x-values.

Solution With $y = 0$, the equation becomes $6x + 2(0) = 15$. Solving for x we obtain $x = 2.5$. When $y = 3$, the equation becomes $6x + 2(3) = 15$, resulting in $x = 1.5$. Continuing in this manner, we generate Table 2.9, all of whose points lie on the graph previously shown in Figure 2.26

TABLE 2.9 Solutions to the Equation $6x + 2y = 15$.

y	0	3	6	9	12
x	2.5	1.5	0.5	−0.5	−1.5

For any linear equation in two quantities, say x and y, solutions can always be found by selecting different values for one of the quantities and then using the equation to determine the corresponding values of the other quantity.[2] By choosing different values of one quantity, we are obviously *varying* or changing the values of that quantity, and that is why the quantities themselves, say x and y, are mathematically referred to as variables.

Graphing a linear equation is significantly simplified because we know that the graph is a straight line, which is uniquely determined by two distinct points. Thus, rather than finding many points on the graph, as we did in the previous three examples, we need find only *two* points and then draw a straight line through them.

---

[2] The equation need not be linear for this approach to work; however, for non-linear equations it is frequently difficult to determine the corresponding value because simple algebraic procedures are not possible. For quadratic equations the quadratic formula can be used, as presented in Section 3.6, if necessary, but such solutions are not available for higher-order equation.

Example 7 Graph the equation $0.5x - 0.75y = 1.7$.

Solution We first note that this equation is a linear equation in the variables x and y; hence its graph is a straight line, As such, it can be easily graphed by finding two points satisfying the equation and then drawing a straight line through these two points. Arbitrarily choosing two values of x, say $x = 0$ and $x = 1$, we find their associated y-values. When $x = 0$, the equation becomes

$$0.5(0) - 0.75y = 1.7$$
$$-0.75y = 1.7$$
$$y = -1.7/0.75 = -2.27$$

When $x = 1$, the equation becomes

$$0.5(1) - 0.75y = 1.7$$
$$-0.75y = 1.7 - 0.5$$
$$-0.75y = 1.2$$
$$y = -1.2/0.75 = -1.6$$

Plotting these two points, $(0, -2.27)$ and $(1, -1.6)$, we obtain Figure 2.28. Then, drawing a straight line through the points, we obtain Figure 2.29 as the graph of the equation

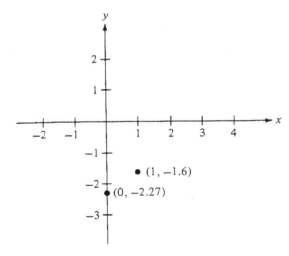

FIGURE 2.28 Plotting two solutions of the equation $0.5x - 0.75y = 1.7$.

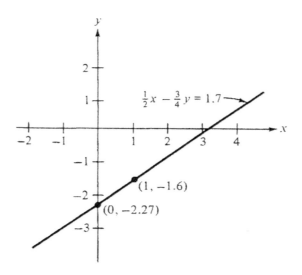

FIGURE 2.29 The graph of the equation $0.5x - 0.75y = 1.7$.

Warning: The method of plotting two points and then drawing a straight line through them is valid *only if* the equation is known to be a linear equation. When applied to equations that are not linear equations, such as $x^2 + y^2 = 4$, this procedure results in erroneous graphs.

Two Special Cases

Finally, we consider two special straight lines given by Equation 2.1 when either A or B equals zero. Such lines can be simplified into the following equations:

$$y = k \qquad \text{(Eq. 2.2)}$$
$$x = h \qquad \text{(Eq. 2.3)}$$

Equation 2.2 is a linear equation whose graph is a horizontal line through the point $x = 0$, $y = k$, which is shown in Figure 2.30, while Equation 2.3 represents a linear equation whose graph is a vertical line through the point $x = h$, $y = 0$, as shown in Figure 2.31.

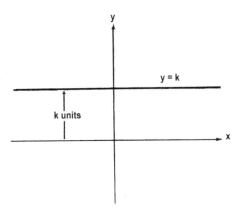

FIGURE 2.30 The graph of the equation $y = k$.

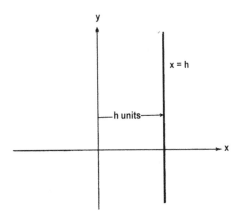

FIGURE 2.31 The graph of the equation $x = h$.

Note that the graph of Equation 2.2, shown in Figure 2.29, is a straight line parallel to the x-axis having all y-coordinates equal to k, while the graph of Equation 2.3, shown in Figure 2.31, is a straight line parallel to the y-axis having all x-coordinates equal to h. What follows from this is that the graph of the line $x = 0$ is the y-axis and the graph of the line $y = 0$ is the x-axis.

Example 8 Graph the equation $y = 5$.

Solution This is an example of Equation 2.2 with $k = 5$. A casual approach might be to assume that x is always zero and $y = 5$, but this would be wrong and almost completely opposite the actual situation.

The equation $y = 5$ provides absolutely no constraints on x, so x *can be any value*. The only way we could conclude that x is zero would be to have the equation $x = 0$, which is not the case here.

From a closer look at the given equation, we see that $y = 5$ can be written as

$$0x + 1y = 5$$

Thus, as long as $y = 5$, any value of x satisfies this equation. In particular, both the points $(1, 5)$ and $(-3, 5)$ satisfy either forms of the equation. Plotting these points and then drawing a straight line through them we find the graph of $y = 5$ is a straight line parallel to the x-axis having all y-coordinates equal to 5 (see Figure 2.32).

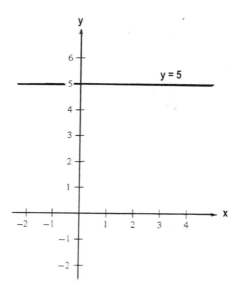

FIGURE 2.32 The graph of the equation $y = 5$.

Exercises 2.3

1. Determine which of the following equations are linear:

   a. $2x = y$ **b.** $2x = 1/y$ **c.** $xy = 4$

   d. $x = 4$ **e.** $2x - 3y = 0$ **f.** $y = 4x$

   g. $y = 4x^2$ **h.** $x - 2 = 3y$ **i.** $1/x + 1/y = 2$

   j. $x + 9 = 4$ **k.** $y - 7 = 0$ **l.** $-2x + 9y = 0$

2. Graph the following equations in the given variables:

   a. $-2p + 3q = -6$ **b.** $-n + 2m = -10$

   c. $n = -1 + 2m$ **d.** $5r - 2s = 0.$

3. Graph the following equations in the given variables:

   a. $2x + 3y = 6$ **b.** $-2x + 3y = 6$ **c.** $2x - 3y = 6$

   d. $2x + 3y = -6$ **e.** $3x + 2y = 6$ **f.** $3y - 4x = 7$

   g. $y = x^3 - 2x^2 + x$ **h.** $x = 7$ **i.** $10x - 5y = 50$

4. The sales (in millions of dollars) of a particular company are given by $S = 02E + 100$, where E represents advertising expenditures (in thousands of dollars). Determine the amount of money that must be committed to advertising in order to realize gross sales of $10 million.

5. After carefully studying used car price lists, Mr. Henry has determined that his particular model car, purchased yesterday for $6000, will depreciate $1,500 per year. Let V denote the value of Mr. Henry's car at any given time t, where t denotes time measured in years. Determine the equation relating t to V and show that it is a linear equation.

6. Benjamin is going to college and needs a new computer. Through his research, he purchased a computer for $2,000. The depreciation rate for this type of computer is 20% per year. Determine the worth of the computer after three and a half years.

7. Pete's Steak House monthly profit, P, is approximated by the equation $P = 250N - 750$, where N represents the number of months. Determine the number of months that it will take for the restaurant to make $5,000.

2.4 PROPERTIES OF STRAIGHT LINES

Linear equations and their corresponding straight lines have a number of extremely useful characteristics. These include a concept called the line's slope, the line's y-intercept, and the ease with which these two quantities can be found from the line's equation and graph. Conversely, the line's equation can easily be determined if two solutions, that is, two points on the line are known. Each of these topics is presented in this section. We begin with the concept of a line's slope.

The Slope

Graphically, a line's slope is the direction and steepness of the line. Mathematically, it provides the rate of change, that is, how fast or slow the y variable changes with respect to a change in the x variable.

Graphically, a line with a positive slope rises as you move from left to right along the x-axis, as shown in Figure 2.33. Also, as shown in the figure, a positively sloped line will have an angle between 0 and 90 degrees between the line and the positively directed x-axis. Finally, a line with a large positive slope, such as 100, is steeper upward (that is, rises more quickly) than a line with a less positive slope, such as 5.

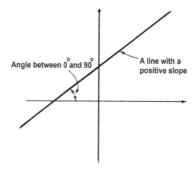

FIGURE 2.33 The graph of a line with a positive slope.

A line with a negative slope falls as you move from left to right along the x-axis, as shown in Figure 2.34. Also, as shown in the figure, a negatively sloped line will have an angle between 90 and 180 degrees between the line and the positively directed x-axis. A line with a large negative slope, such as −100, is steeper downward (that is, falls more quickly) than a line with a less negative slope, such as −5.

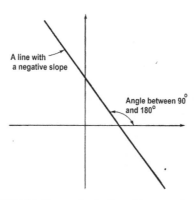

FIGURE 2.34 The graph of a line with a negative slope.

A line parallel to the x-axis, as previously illustrated in Figure 2.30, has a slope of 0, while a line parallel to the y-axis, as previously illustrated in Figure 2.31 does not have a slope. Mathematically, the slope of a straight line is defined as follows:

Definition 2.2 Let (x_1, y_1) and (x_2, y_2) be any two distinct points on the same straight line (or, alternatively, two solutions to the same linear equation).

If $x_2 - x_1$ *is not* zero, then the *slope* of the line, denoted as m, is:

$$m = slope = \frac{y_2 - y_1}{x_2 - x_1} \qquad \text{(Eq. 2.4)}$$

providing $x_2 \neq x_1$. If $x_2 - x_1$ is zero, the slope is undefined.

The numerator of Equation 2.4, $(y_2 - y_1)$ represents the vertical distance between the two points (x_1, y_1) and (x_2, y_2) and is formally referred to as the *rise*, while the value of the denominator $(x_2 - x_1)$ is formally referred to as *run*. Using these two terms, an alternative form of Eq. 2.4 is:

$$m = slope = \frac{rise}{run} \qquad \text{(Eq. 2.4a)}$$

Again, the slope is undefined if the run is zero.

Either Equation 2.4 or 2.4a provides a formula for calculating the slope of a straight line using any two points on the line. To find the slope, we first must subtract y-values, which determines the value of the rise, subtract their corresponding x-values in the same order as the y-values were subtracted, which determines the value of the run, and then divide the subtracted y-values by the subtracted x-values. The result obtained is the *slope*.

Thus, to find the slope of a given equation:

- *Choose a value for x*
- *Substitute the value of x into the given equation and solve the equation for its corresponding y*
- *Repeat this process for the next x value*
- *Use the formula $m = (y_2 - y_1)/(x_2 - x_1)$, or alternatively $m = rise/run$ to find the slope of the given equation.*

Example 1 Find the slope of the line $6x + 2y = 15$.

Solution In Example 1 in Section 2.2 we found that two points on this line are $(0, 7.5)$ and $(1, 4.5)$. Therefore, letting $x_1 = 0$, $y_1 = 7.5$, and $x_2 = 1$, $y_2 = 4.5$ and substituting into Equation 2.10, we have

$$Slope = \frac{4.5 - 7.5}{1 - 0} = \frac{-3}{1} = -3.$$

Notice that the same value of the slope would be obtained if we had selected $(1, 4.5)$ as the (x_1, y_1) point and $(0, 7.5)$ as the (x_2, y_2) point, or any other two points on the line. Definition 2.2 simply requires that *the x-values of the two points be subtracted in the same order as their corresponding y-values*.

Example 2 Find the slope of the linear equation $y - 2x = 1$.

Solution Definition 2.2 requires that we select two points satisfying the given equation $y - 2x = 1$. Arbitrarily selecting x-values of 1 and 4 (any choice would do) we calculate the corresponding y-values as 3 and 9. Thus two points satisfying the equation are $(1, 3)$ and $(4, 9)$. Letting $(1, 3)$ be the (x_1, y_1) point, $(4, 9)$ as the (x_2, y_2) point, and substituting into Equation 2.3 yields:

$$m = Slope = \frac{9 - 3}{4 - 1} = \frac{6}{3} = 2.$$

Note that a line parallel to the x-axis, such as the one previously shown in Figure 2.29 has a slope of zero. The reason is that any two points on the line will have the same y value. Substituting this value into Equation 2.4 results in a numerator value of zero, which results in a zero slope.

Similarly, a line parallel to the y-axis, such as the one previously shown in Figure 2.31 does not have a slope. Here, the reason is that any two points on such a line have the same x value. Substituting this value into Equation 2.4 results in a denominator value of zero, which is mathematically undefined.

The Slope as a Rate of Change

The two points used to determine a line's slope can be any randomly selected distinct points, as shown in Figure 2.35 as the two points P_1 and P_2. As indicated, the difference in the y-values, $y_2 - y_1$, represents the vertical distance between the two points P_1 and P_2, while the difference in the x-values represents the horizontal distance between the two points.

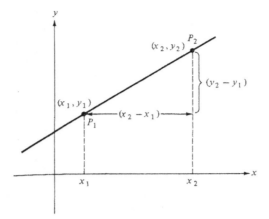

FIGURE 2.35 The change in x and y-values.

A specific example should clarify this point. In Example 2, we showed that the line satisfying the equation $y - 2x = 1$ had a slope of 2. This curve, along with the two points P_1 and P_2 used to calculate the slope are shown in Figure 2.36.

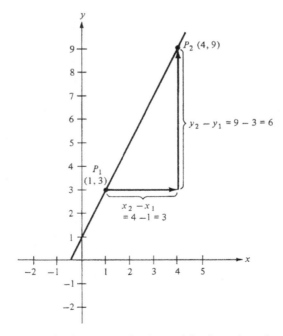

FIGURE 2.36 The change in x-values is 3 and the change in y-values is 6.

From the graph it is evident that, had we started at the point (1, 3) and moved to the point (4, 9) along the line, we would have traveled a distance of 3 units to the right and up a distance of 6 units, as indicated by the arrows. The 6 and 3 are the respective differences in y- and x-values from point P_1 to point P_2. The ratio 6/3, or simply 2, is the slope of the line.

Recall that we selected points P_1 and point P_2 in Figure 2.36 at random. Had we selected two other points on the curve and calculated the slope, we still would have found the slope to be 2. The reason for this is that any two points on the curve illustrated in Figure 2.36 are related in a particular way, as follows: by starting at one point on the curve, every other point on the line can be reached by increasing or decreasing y- and x-values in the ratio of 2 to 1. If we start at point P_1 and increase x by 1 unit, we must increase y by 2 units to land back on the line. Should we increase x by 5 units, we must increase y by 10 units to remain on the line.

The significance of the slope is that it gives us the particular ratio relating all points on a given line. As such, ***the slope represents the rate of change in y associated with a change in x***, and this has both interesting and important commercial applications.

Example 3 It is known that the monthly sales of a particular model automobile S (in units) is related to the advertising expenditures E (in millions of dollars) by the equation $S = 20,000 + 5,000E$. Determine the rate of change in sales with respect to advertising expenditures.

Solution If we rewrite the given equation as $S - 5,000E = 20,000$, we observe it conforms to the definition of a linear equation; as such, its graph is a straight line. The rate of change is just the slope. To find the slope, we first need two distinct points on the line. Arbitrarily choosing $E_1 = 0$ and $E_2 = 1$ (any two values could have been selected), we find the corresponding values of S as $S_1 = 20,000$ and $S_2 = 25,000$. The rate of change in sales with respect to advertising expenditures is

$$Rate\ of\ Change = m = Slope = \frac{25,000 - 20,000}{1 - 0} = 5,000$$

This means that whenever E is increased by 1 unit (in this case \$1 million), the monthly sales will be increased by 5,000 units (in this case cars).

Slope-Intercept Form

Recall from Definition 2.1 that a linear equation has the form $Ax + By = C$. For any non-vertical straight line a frequently useful alternate form of this equation is[3]:

$$\boldsymbol{y = mx + b} \tag{Eq. 2.5}$$

where m is the line's slope and b is known as the *y-intercept*. The reason for this name is that the point $(0, b)$ is where the line crosses the y-axis, as shown in Figure 2.37.

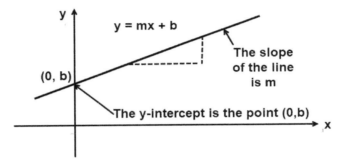

FIGURE 2.37 The slope and *y*-intercept of a straight line.

Equation 2.5 is known as *the slope-intercept form* of a linear equation and is an alternative representation to Equation 2.1 for all straight lines not parallel to the y-axis. For lines parallel to the y-axis the slope is undefined and Equation 2.5 is not valid. For all other straight lines, however, Equation 2.5 is valid, and has an advantages over Equation 2.1 in that the slope, m, is immediately visible in the equation as the coefficient of the x term.

Example 4 Determine the slope of the line $4y - 12x = 36$

Solution Rewriting the equation into the form $y = mx + b$, we have

$$4y = 12x + 36$$
$$y = 3x + 9$$

Once the equation is in this form, the coefficient in front of the x term yields the slope. Thus, the slope of the line is 3.

---

[3] As the slope of a vertical line, such as that shown in Figure 2.26, is undefined, Equation 2.5 cannot be used for these lines.

The value b in Equation 2.5 is the value at which the line crosses the y-axis, and is referred to as the *y-intercept.*

Example 5 Find the slope and the y-intercept of the line $3x + 4y = 10$.

Solution Solving the equation for y, results in the slope-intercept form of the equation as

$$y = -0.75x + 2.5$$

with $m = -0.75$ and $b = 2.5$. Therefore, the slope is -0.75 and the y-intercept is located at $(0, 2.5)$.

Point-Slope Form

The Slope-Intercept form of a linear equation (Equation 2.5) is most useful for determining a line's slope and y-intercept from either its equation or graph. However, the reverse situation of determining a line's equation when two points known to be on the line are provided, is not as easily obtained. For this situation, a more useful form of a linear equation is

$$\boldsymbol{y - y_1 = m(x - x_1)} \qquad \text{(Eq. 2.6)}$$

where (x_1, y_1) and (x_2, y_2) are two points on the line and

$$\boldsymbol{m = slope = \frac{y_2 - y_1}{x_2 - x_1}.}$$

Equation 2.6 is known as *the slope-intercept form* of a linear equation.

Example 6 Find the equation of the straight line passing through the points $(3, 15)$ and $(-4, 50)$.

Solution Because these points are on the same straight line, the slope of the line can be computed as

$$\text{Slope} = m = \frac{y_2 - y_1}{x_2 - x_1} = \frac{50 - 15}{-4 - 3} = \frac{35}{-7} = -5.$$

From Equation 2.16 the line will have the form $y - y_1 = -5(x - x_1)$. Substituting 3 for x_1 and 15 for y_1 the linear equation containing the two given points is

$$y - 15 = -5\,(x - 3)$$

This equation can be rewritten as $y = -5x + 30$, which is the slope-intercept form of the equation.

Example 7 Table 2.10 is the result of years of data collecting by the American Citrus Corporation. Using this data, determine the equation relating the number of orange trees that bear fruit, F, to the number of trees planted N.

TABLE 2.10

N (number of trees planted)	120	140	160	180
F (number that bear fruit)	114	133	152	171

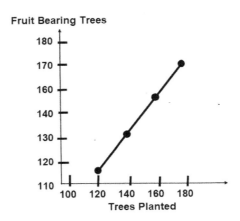

FIGURE 2.38 Fruit bearing trees versus trees planted.

Solution The four points listed in Table 2.10 are plotted in Figure 2.38. As they fall on a straight line, the variables F and N are related by a linear equation. Two points on the line are given by the first two points in Table 2.10 (any two points could be used) as (120, 114) and (140, 133). Thus, the slope of the line is

$$\text{Slope} = m = \frac{F_2 - F_1}{N_2 - N_1} = \frac{133 - 114}{140 - 120} = \frac{19}{20} = 0.95$$

Using the point-slope equation for a line with the letters F and N replacing y and x, respectively, yields

$$F - 114 = 0.95(N - 120)$$

which, simplified becomes $F = 0.95N$. This is the equation of a straight line through the origin, even though that fact is not apparent from the section of the graph displayed in Figure 2.38.

Exercises 2.4

1. Find the slopes of the following straight lines:

   a. $2x + 3y = 6$ **b.** $-2x + 3y = 6$ **c.** $2x - 3y = 6$

   d. $2x + 3y = -6$ **e.** $3x + 2y = 6$ **f.** $x = 7$

   g. $10x - 5y = 50$ **h.** $x = y.$ **i.** $7x - 2y = 14$

   j. $x - 3y = 9$ **k.** $9x + 2y = 18$ **l.** $-2x + y = -4$

2. Find the equation of the straight line containing the given points:

   a. $(1, 2)$ and $(2, 5)$ **b.** $(7, -3)$ and $(-1, -8)$

   c. $(-1, 2)$ and $(4, 2)$ **d.** $(1, 0)$ and $(0, 1)$

   e. $(2, -1)$ and $(2, 4)$ **f.** $(-1, 4)$ and $(0, 2)$

   g. $(7, 5)$ and $(9, -3)$ **h.** $(5, -1)$ and $(1, 11)$

3. Figure 2.39 illustrates the cumulative monthly attendance at a local amusement park for the past year. Determine the equation relating attendance, A, to time t (in months), assuming $t = 0$ corresponds to January.

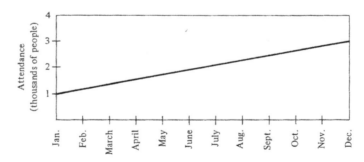

FIGURE 2.39 Through 2.41 do not need captions, as they are used in the Exercises.

4. Figure 2.40 illustrates the cumulative weekly sales receipts of a supermarket over the past year.

   a. Determine the equation relating gross income, I, to time, t, for the first 20-week period.

   b. Determine the equation relating I to t for the last 32 weeks of the year.

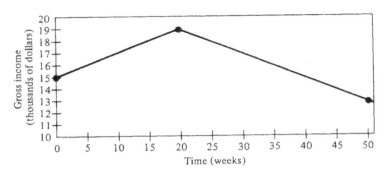

FIGURE 2.40

5. After test-marketing a new bleach, the White-All Bleach Company collected the data are given in Table 2.11. Plot the points in Table 2.6 to verify that the relationship between the number of subsequent purchases P and the number of samples distributed S is a straight line, and then determine the equation of that line.

TABLE 2.11

N (number of free samples)	1000	1500	2000	3000	5000
P (number of subsequent purchases)	2050	2075	2100	2150	2250

6. Quality-control tests on the manufacture of light bulbs resulted in Table 2.12. Plot the points in this table to verify that the relationship between the number of defective bulbs, D and the number of bulbs produced N is a straight line, and then determine the equation of that line.

TABLE 2.12

N (number of bulbs produced)	25,000	50,000	60,000	75,000	90,000
D (number of defective bulbs)	75	150	180	225	270

7. At a recent sales meeting of the Lincoln Hamburger Company, Figure 2.41 was displayed to emphasize the growth in profit. Find the equation relating profit, P, to time t (in years).

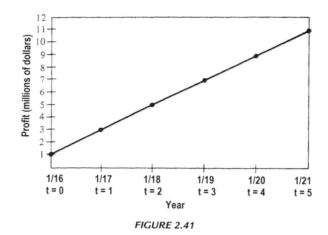

FIGURE 2.41

8. A large television manufacturer has determined that the number of television sets sold, denoted by N, is related directly to the amount of money spent on advertising. In particular, every million dollars in advertising expenditures results in an additional 50,000 television sets being sold, although 10,000 sets would be sold with no advertising. Let E denote the amount of money (in millions of dollars) committed to advertising. Determine the equation relating N to E and show that it is a linear equation.

9. An appliance salesperson working for a high-end department store determines that the number of each appliance sold, denoted by x, is directly related to the income she earns each year, denoted by y. Her base salary is $35,000 and, for each appliance she sold, she will earn a commission of $250.

   a. Determine the equation relating x to y and show that this equation is linear.

   b. What would be the salesperson's income if she sold 70 appliances by the end of the year?

   c. How many appliances would the salesperson need to sell to earn a $75,000 income?

2.5 BREAK-EVEN ANALYSIS

Linear equations are extremely useful in business applications for determining the relationship between short-term revenue and short-term costs. Conventionally, the term *short-term* refers to a time period in which both the price and the cost of an item remain constant. Over more extended time periods, economic conditions, such as inflation, supply and demand, and other economic factors typically act to change the cost and price structures. Over the short term, which is generally defined as a year or less, these other factors tend to have little direct influence.

The **Break-Even point** is the point at which the income from the sale of manufactured or purchased items exactly matches the cost of the items being sold. When this happens, the seller neither makes nor loses money but simply breaks even.

The reason the break-even point is so important is that it provides business information about the sales at which the company switches over from incurring a loss to making a profit. Should it be decided that sales can be higher than the break-even point, it means a profit can be made; otherwise, any sales less than the break-even point indicates that the venture will result in a loss. As such, it also forms as a lower bound for marketing, because if the break-even point cannot be reached, spending time and effort in marketing becomes a futile endeavor.

In determining the break-even point, both the revenue obtained by selling, and the cost involved in acquiring the items being sold must be taken into account.

The Revenue Equation

By definition, **revenue** is the income obtained from selling goods or services. In its simplest form, the revenue produced from a sale, known as the *sales revenue*, is simply the price of each item times the number of items sold.[4] Designating the sales revenue by R, the price per unit by p and the number of items sold by x, we have

$$R = px, \tag{Eq. 2.12}$$

Because p is assumed fixed and known, Equation 2.12 is a linear equation in the variables R and x.

Example 1 A company that manufactures calculators has a contract to sell calculators for $5.00 a piece to a discount electronics outlet chain. Determine the revenue equation and the actual revenue realized if 2,000 calculators are sold.

Solution Using Equation 2.12 the revenue equation is

$$R = \$5.00\, x$$

If 2,000 calculators are sold, the revenue, R, realized is

$$R = \$5.00(2,000) = \$10,000.$$

---

[4] Here, we are restricting ourselves to the short run (typically defined as a year or less) in which the price of each item does not change.

The Cost Equation

The cost of items sold are commonly separated into two categories: fixed costs and variable costs.

Fixed costs include rent, insurance, property taxes, and other expenses that are present regardless of the number of items produced or purchased. Over the short run these costs are fixed because they exist and must be paid even if no items are purchased for resale or produced and sold. We will represent the fixed cost by the variable F.

Variable costs are those expenses that are directly attributable to the manufacture or purchase of the items themselves, such as labor and raw materials. Variable costs depend directly on the number of items manufactured or purchased – the more items manufactured or purchased, the higher the variable costs. If we restrict ourselves to short-run conditions, the cost-per-item is a fixed number, which makes the variable cost equal to this cost-per-item times the number of items purchased or manufactured. Designating the variable cost by V, the cost-per-item by a, and the number of items manufactured or purchased by x, we have

$$V = ax \qquad \text{(Eq. 2.13)}$$

Because the total cost is the sum of the variable cost plus fixed cost, the total cost equation becomes

$$C = V + F \qquad \text{(Eq. 2.14)}$$

Substituting Equation 2.13 for V into equation 2.14, the final cost equation becomes

$$C = ax + F \qquad \text{(Eq. 2.15)}$$

That is, the total cost is the sum of the variable cost and the fixed cost. The numbers a and F are assumed known and fixed; hence Equation 2.15 is a linear equation in C and x.

Example 2 A company manufacturing electronic calculators have recently signed contracts with its suppliers. For the duration of these contracts, the cost of manufacturing each calculator is $1.20. The company estimates that the fixed costs for this period will be $8,000. Determine the total cost function for this process and the actual cost incurred if only 500 calculators are actually manufactured.

Solution Using Equation 2.15 with $a = 1.20$ and $F = \$8,000$, we have

$$C = \$1.20x + \$8,000.$$

If 500 calculators are produced, the cost will be $C = \$1.20(500) + \$8,000 = \$8,600$. If no calculators are produced, the total cost will be $C = \$1.20(0) + \$8,000$ or $\$8,000$, which is the fixed cost.

The Break-Even Point

From Examples 1 and 2, we note that a production run of 500 calculators will result in a total cost of $8,600 and a sales revenue of only $2,500. The company will experience a loss of $6,100. Such embarrassing situations can be avoided with a *break-even analysis*. As the name suggests, this analysis involves finding the level of sales below which it will be unprofitable to produce items and above which sales revenue exceeds costs so that a profit is made. This level is the *break-even point*. The break-even point occurs when total cost exactly equals sales revenue.

If we restrict ourselves to the short run and assume that all items produced can be sold, the break-even point is obtained by setting the right side of Equation 2.12 equal to the right side of Equation 2.15. That is, the break-even point occurs when $R = C$. Substituting for both the revenue, R, and cost, C, from Equations 2.12 and 2.15 yields

$$px = ax + F \qquad \text{(Eq. 2.16)}$$

Equation 2.16 is one equation in the one unknown, x. Solving for x using the algebraic methods presented in Section 1.2 yields the break-even point, *BEP*, as

$$\textbf{BEP} = \textbf{\textit{x}} = \textbf{\textit{F}} \textbf{/} \textbf{(\textit{p} − \textit{a})} \qquad \text{(Eq. 2.17)}$$

For the electronic calculator described in Examples 1 and 2, we found $C = \$1.20x + \$8,000$ and $R = \$5.00x$. The BEP occurs when $R = C$, or, from Equation 2.17, when $x = 8,000/(5.00 − 1.20) = 2,106$ calculators. Any production and sales below 2,106 calculators results in a loss, while any production and sales above 2,106 units produces a profit.

Example 3 A lamp component manufacturer determines that the manufacturing costs associated with each component are $5 and that the fixed costs are $7,000. Determine the BEP if each component sells for $7. Assume that each unit made can be sold.

Solution The total cost for this process, using Equation 2.15, is $C = \$5x + \$7,000$. The sales revenue is $R = \$7x$. The BEP is the value of x for which $R = C$. This point can be found by directly using Equation 2.17, which yields, $x = 7000/(7 − 5) = 3,500$ components as the BEP.

Example 4 A dress manufacturer determines that the production costs associated directly with each dress are $8 and that the fixed costs are $9,200. Determine the BEP if each dress sells for $54. Assume that all dresses manufactured can be sold.

Solution The total cost for this process is given by Equation 2.15 as $C = \$8x + \$9,200$. The sales revenue is given by Equation 2.12 as $R = \$54x$. The BEP is the solution of the equation $x = 9,200/(54 - 8)$, which yields $x = 200$ dresses.

Graphical Solutions

Break-even problems can be solved graphically as well as algebraically. The procedure is to plot both the revenue and cost equations on the same graph, as shown in Figure 2.42. Because both of these equations are linear, their graphs will both be straight lines.

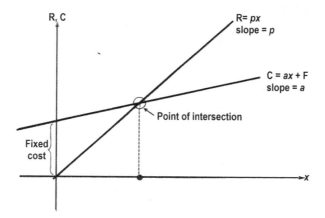

FIGURE 2.42 The graph of linear revenue and cost equations.

In reviewing Figure 2.42, note that the horizontal axis is x, the number of units produced and sold, whereas the vertical axis is R and C, depending on which equation is being considered. It follows that the unit sales price p is the slope of the revenue equation (defined by Equation 2.12), and the cost of manufacturing each unit, denoted as a, is the slope of the cost equation (defined by Equation 2.15). The y-intercept corresponding to Equation 2.15 is simply the fixed cost.

The break-even point (BEP) is the value of x for which $R = C$, which in Figure 2.42 is the value of x at the intersection point of the two lines.

Example 5 Graphically determine the BEP for the manufacturing process described in Example 3.

Solution In Example 3, we determined the equations $C = 5x + 7,000$ and $R = 7x$. Using the graphing procedures given in Section 2.2, we plot each line on the same graph, as shown in Figure 2.43. The x-component of the intersection point of these two lines is read directly from the graph as $x = 3,500$, which is the same BEP found algebraically in Example 3.

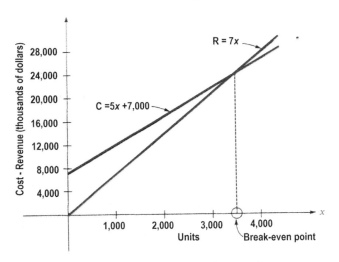

FIGURE 2.43 Locating the Break-even point.

Exercises 2.5

1. A publisher of a current economics textbook determines that the manufacturing costs directly attributable to each book are $55 and that the fixed costs of production are $225,000. The publisher sells each book for $100 per copy.

   a. Determine the equation relating the total cost to the number of books published.

   b. Determine the equation relating the sales revenue to the number of books sold.

   c. Determine the profit if 3000 books are published and sold.

   d. Algebraically determine the BEP for this process.

2. Determine the BEP in Exercise 1 graphically. Which method do you prefer?

3. A manufacturer of staplers determines that the variable costs directly attributable to each stapler are $2 and that the fixed costs are $15,000. Each stapler sells for $12.00. Determine the BEP for this process both graphically and algebraically.

4. Using the results from Exercise 3, determine

   a. the total cost of the process at the BEP

   b. the total sales revenue of the process at the BEP

   c. the profit at the BEP.

5. A manufacturer of light bulbs determines that each bulb costs 15¢ in direct expenses and that the process as a whole incurs fixed costs of $15,000.

   a. Determine the BEP if each bulb sells for $1.75.

   b. Determine the BEP jf each bulb sells for $2.50.

   c. Does it make sense that the answer in part b. is smaller than that in part a.?

6. A manufacturer of Lucite pipe holders has determined that the firm has a BEP of 2,500 units. Determine the price of each holder if each item costs $3.00 to manufacture and the process involves fixed costs of $15,000.

7. A manufacturer of specialty bookends has determined that the BEP for the manufacturing process is 120 units. Determine the variable cost of producing each bookend set, if the fixed costs are $1,920 and each set sells for $25.

8. A publisher of a current solution manual to a textbook determines that the manufacturing costs directly attributed to each manual are $4 and that the fixed costs are $15,000. The publisher sells each manual for $29.99 per copy.

   a. Determine the equation relating the sales revenue to the number of manuals published.

   b. Determine the equation relating the total cost to the number of manuals.

   c. What will the profit be if 2000 manuals are published?

   d. Using Equation 2.17 determine the BEP for this process.

9. Determine the BEP in Exercise 8 graphically. Which method do you prefer?

10. A manufacturer of calculators determines that the variable costs directly attributable to each calculator is $25 and that the fixed costs are $20,000. Each calculator sells for $99.

    a. Determine the BEP for this process both graphically and algebraically.

    b. Determine the total sales revenue of the process at the BEP.

    c. Determine the total cost of the process at the BEP

    d. Determine the profit at the BEP.

2.6 CONSTRUCTING LINE GRAPHS USING EXCEL[5]

The majority of line graphs, such as the one shown in Figure 2.44, present the relationship between two variables, such as Sales versus Advertising, or between a quantity, such as revenue or profit, as it changes over time, such as years in business. Either straight or curved lines can be used to connect individual points on the graph.[6]

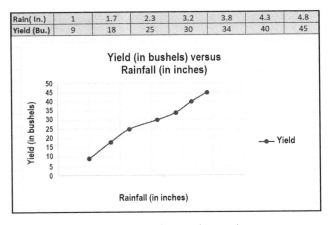

Rain(In.)	1	1.7	2.3	3.2	3.8	4.3	4.8
Yield (Bu.)	9	18	25	30	34	40	45

FIGURE 2.44 A line graph example.

---

[5] This section may be read independently of any other sections in this chapter.

[6] A line graph without connecting lines between data points is a scatter diagram. Scatter diagrams are typically used with trend lines, as presented in the section 7.3.

Creating a line graph requires that the data first be entered into either consecutive rows, as shown in the upper-left corner of Figure 2.44 or consecutive columns. Once entered, the data is highlighted, as shown as Step 1 in Figure 2.45, and the remaining steps shown in the figure then be completed.

Although any two consecutive rows or columns can be used for the data, when rows are used, the data entered in the upper-most row become the x-axis values, and the data in the lower row become the y-axis values. Similarly, for data listed in columns, the data in the left-most column is plotted on the x-axis and the data in the right-most column on the y-axis.

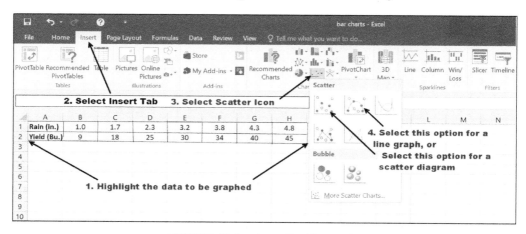

FIGURE 2.45 Creating an Excel line graph.

After selecting the line graph option shown as Step 4 in Figure 2.42, the line graph shown in Figure 2.46 is automatically created. Notice that the figure has no axes titles and that the chart's title is a copy of the data label in cell A2, which is the y-axis data label.

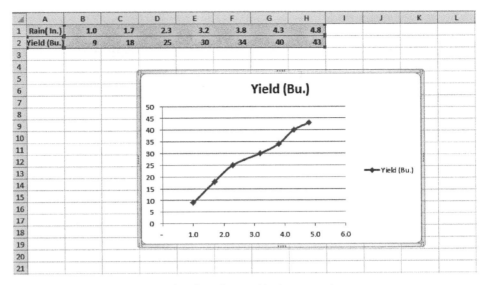

◢	A	B	C	D	E	F	G	H	I	J	K	L
1	Rain(In.)	1.0	1.7	2.3	3.2	3.8	4.3	4.8				
2	Yield (Bu.)	9	18	25	30	34	40	43				

FIGURE 2.46 Changing a line graph's elements and appearance.

Figure 2.47 shows the initial position and description of the four descriptive elements associated with line graphs. These consist of a chart title, two axes labels and a legend. When any of these textual elements are present the text, font type and size of each can be changed by clicking on the desired element and then entering new text and font attributes.

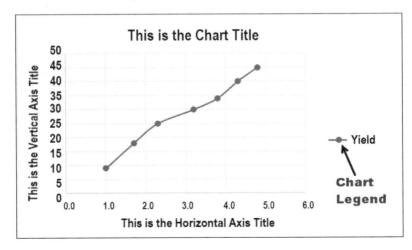

FIGURE 2.47

Deleting or adding a chart element is accomplished using the Chart Elements submenu shown in Figure 2.48. As shown in the figure, this submenu is activated by either right- or left-clicking within the chart, and then clicking on the cross-hairs icon and selecting the desired element.

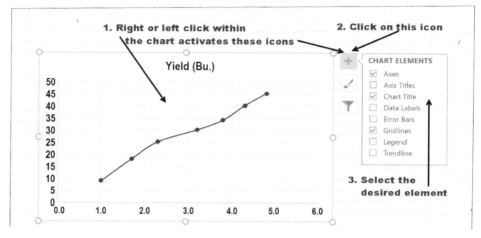

FIGURE 2.48 Adding line graph titles and axes labels.

Additionally, a line graph's legends and line colors can be changed using the Change Colors option in the Chart Tools Design ribbon, shown in Figure 2.49. The Design ribbon is obtained by either double-clicking within the chart or single-clicking and then selecting the Chart tool's Design tab (seen at the top of Figure 2.49).

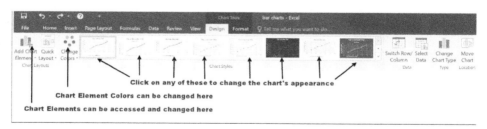

FIGURE 2.49 The chart design ribbon.

Finally, a line graph's appearance is changed by selecting one of the chart styles shown within the Chart Design ribbon (again, see Figure 2.49). Clicking on any of the pre-constructed styles changes the chart to the selected choice.[7]

---

[7] Notice that chart elements previously shown in Figure 2.48 can also be accessed and changed by clicking on the leftmost icon in the Chart Design ribbon.

Changing Chart Types

As with all Excel charts, a chart can easily be changed to a different chart type. Figure 2.50 illustrates two variations for the line graph previously illustrated in Figure 2.48.

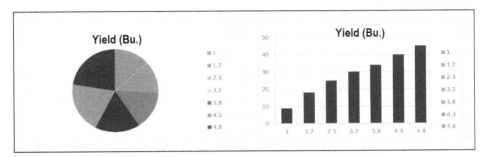

FIGURE 2.50 Different chart types.

As with all of Excel's charts, a line graph can be changed to another type by:

1. Right-clicking within the chart.
2. Clicking on the Change Chart Type option.
3. Selecting the desired chart type from the resulting pop-up submenu.

The first two steps are illustrated in Figure 2.51.

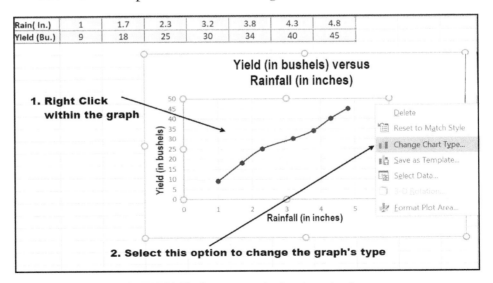

FIGURE 2.51 The first two steps in changing a chart's type.

In addition to creating a graph with a single line, Excel permits multiple lines to be plotted on the same graph. The multiple line graph shown in Figure 2.52 was created using the same technique as presented above, with the exception that all of the data in rows 1, 2, and 3 were highlighted before selecting the line graph option.

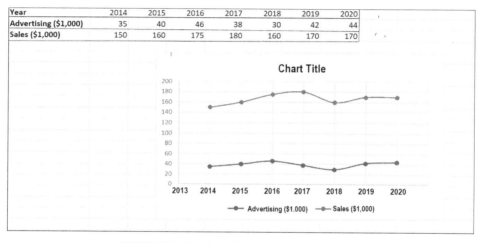

Year	2014	2015	2016	2017	2018	2019	2020
Advertising ($1,000)	35	40	46	38	30	42	44
Sales ($1,000)	150	160	175	180	160	170	170

FIGURE 2.52 Creating a line graph with multiple lines.

2.7   SUMMARY OF KEY POINTS

Key Terms

- Bar Graphs
- Break-even analysis
- Break-even point
- Cartesian coordinated system
- Coordinates
- Fixed costs
- Graph of an equation
- Horizontal intercept
- Line Graphs
- Linear equation
- Origin
- Pie Chart

- Piecewise Graphs
- Revenue
- Rise
- Run
- Scale
- Slope
- Solution
- Straight line
- Tick marks
- x-Coordinate
- y-Coordinate
- Variable costs

Key Concepts

- Graphs have a visual impact and are a powerful medium for presenting data and relationships.
- Every point on a Cartesian coordinate system is uniquely determined by an ordered pair of numbers. By convention, the first number is a value on the horizontal axis, while the second number is a value on the vertical axis.
- Successive tick marks must be equally spaced, and the units between equally spaced tick marks must be the same. However, the scale on the horizontal axis can differ from the sale on the vertical axis.
- Geometrically, the set of all solutions of one equation in two unknowns is a curve on the plane
- Graphing too many points is always better than graphing two few.
- The graph of a linear equation in two variables is a straight line.
- The slope of a straight line is the same regardless of which two points on the line are used to calculate the slope.
- Starting from any point on a line, the slope is the change in the vertical direction (the rise) needed to accompany a unit change in the positive horizontal direction (the run) if the resulting point is to be on the line.
- A line with a positive slope slants upward to the right, while a line with a negative slope slants downward to the right.
- A line parallel to the x-axis has a zero slope; a line parallel to the y-axis does not have a defined slope.

CHAPTER **3**

FUNCTIONS

In this Chapter

Linear equations, which were presented in Chapter 2, are a special case of a broader class of equations known as polynomials. Even though some of the most important business applications can be modeled using linear equations, there are other applications, notably in statistics and marketing that require these more complex polynomials. Additionally, the basics of most financial applications rely on an entirely different class of equations, known as exponentials. A key concept underlying both polynomial and exponential equations, as well as almost all other equations used in business, is a function.

This chapter begins with describing what a function is and how it relates to mathematical equations. This is followed by a detailed presentation of polynomial functions, quadratic functions, which, like linear equations are a particular type of polynomial, and exponential functions.

3.1 CONCEPT OF A FUNCTION

When mathematics is used to model real-world conditions, it must account for relationships between quantities. For example, the growth of bacteria over time, the amount of sales that follow from a given advertising budget, or the increase in the amount earned when a sum of money is invested in a bond.

As an example, Table 3.1 illustrates a relationship between the number of cars sold over time, from 2016 to 2021 by Village Distributors, a small new car dealer. Each year's sales are arranged under the corresponding year, which clearly shows the relationship between the two quantities, year and number of cars' sold.

TABLE 3.1 Yearly Car Sales by Village Distributors.

Year	2016	2017	2018	2019	2020	2021
Cars Sold	160	145	155	102	95	151

Now consider Report 3.1, which is both wordier than Table 3.1 and less useful. The reason is that a clear and direct assignment between individual years and the number of cars sold during each year is not immediately evident.

Report 3.1

Summary of new car sales:

During the years 2016 through 2021, Village Distributors had a spotty sales record. On three occasions car sales went over the 130 mark (151, 155, and 160), but during other years they fell under 150, twice drastically (95 and 102) and once marginally (145).

The notion of two distinct sets of quantities (like years and number of cars sold) and a rule of assignment between the sets, as presented in Table 3.1 by arranging corresponding entries under each other, is central to the concept of a function. In fact, it describes a function.

Definition 3.1 *A function* is an assignment rule between two sets, which assigns to each element in the first set exactly one element (but not necessarily a different one) in the second set.

A function therefore has three components: (1) a first set (perhaps years), (2) a second set (perhaps numbers), and (3) an assignment rule between the two sets. This rule must be complete in that an assignment must be made to each and every element of the first set. As an example, take the first set to be all the people in the world, the second set as all positive numbers, and use the rule, "Assign to each person his or her exact weight." This is a function. We have two sets and a rule which assigns to every element in the first set (people) exactly one element in the second set (his or her weight).

As a second example, let the first set be all the cars in the world, the second set the all the colors, and use the assignment rule, "Assign to each car its color." This is *not* a function. Although we have two valid sets, the rule cannot handle a car with a red body and a white roof. Such a car must be assigned *two* colors, red and white, and this is not valid for function rules. For a function, the assignment rule must assign exactly *one* element of the second set to each element of the first set.

It can be useful to visualize a function as a robotic machine. The machine is programmed to:

1. Accept elements from the first set,

2. Transform these elements according to the rule it has been given, and

3. Output the result, which is an element of the second set.

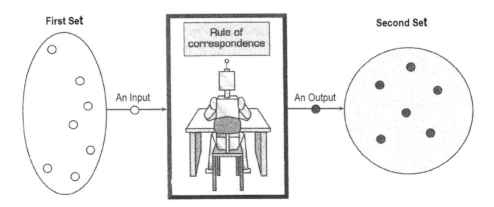

FIGURE 3.1 An assignment rule as represented by a robot.

Note that the robotic machine is incapable of thought. If it receives an input for which the rule does not apply, it will break down. Functions act similarly. The assignment rule must be capable of matching each element of the first set to one element of the second set. If the assignment rule cannot do this, it is not a legitimate function.

In the previous people-weight example, we feed the machine a person, and it emits a number, that person's weight. The machine can handle any person we provide to it. In the car-color example, however, we input a car into the machine. If the car has two colors, the machine will have trouble. Because it cannot think, it will not know which color to assign to a two-tone car, for example, a red and white car.

The people-weight example is interesting in another respect. Although each person is assigned one number, all the assigned numbers are not necessarily different. Two different people can have the same weight. Nonetheless, we still have a function. Definition 3.1 requires only that each element of the first set be assigned one element of the second set, but *not necessarily* a *different one.*

It is too wordy to refer constantly to the two sets under consideration as the first set and the second set. More commonly, the first set is referred to as the *domain*, and the second set is referred to as the *range.*

Example 1 Determine whether or not the two sets of numbers displayed in Figure 3.2 and the assignment illustrated by the arrows is a function.

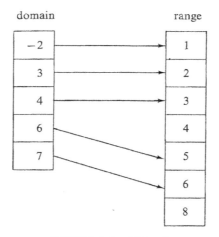

FIGURE 3.2 A valid function.

Solution Because each element of the domain has exactly one element of the range assigned to it, the relationship indicated is a function. Although some elements in the second set are not paired with elements in the first set, this does not matter. Two sets are given, and every element in the first set has assigned to it one element in the second set.

Example 2 Determine whether or not the two sets of numbers displayed in Figure 3.3 and the assignment illustrated by the arrows is a function.

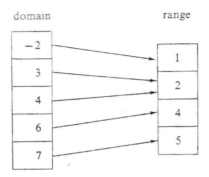

FIGURE 3.3 A valid function.

Solution Because all the conditions specified in Definition 3.1 are satisfied (we have two sets, and each element in the domain is assigned one element in the range), this relationship is a function. As in the people-weight example, we considered previously, some elements of the range are assigned twice. Again, this is of no consequence.

Example 3 Determine whether or not the relationship illustrated in Figure 3.4 is a function.

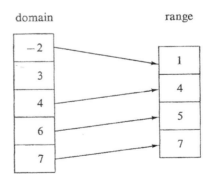

FIGURE 3.4 Not a function.

Solution This relationship is not a function because one element in the domain is not matched with any element in the range. Definition 3.1 requires that each and every element of the first set be assigned an element in the range.

The importance of clearly designating which set is the domain and which is the range cannot be underestimated. To illustrate the pitfalls involved in interchanging their roles, we return to the function defined in Example 2 and reverse the roles of the two sets. Because assignment rules act on the elements in the domain, we also reverse the direction of the arrows. The results are illustrated in Figure 3.5. The relationship given in Figure 3.5 is not a function. Here element 2 in the domain has *two* elements in the range assigned to it. Definition 3.1 requires that each element in the domain be assigned *one and only one* element in the range.

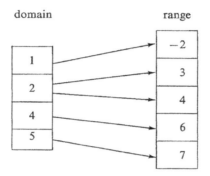

FIGURE 3.5 Not a function.

The selection of which set is to be called the domain and which set is to be called the range is left to the discretion of the person defining the relationship. In general, however, the decision is based on the context of the application. Only after the selection is made and the the assignment rule is given can a decision be made as to whether or not the components form a function.

Exercises 3.1

In Exercises 1 through 14, determine whether or not the given relationships are functions.

1.

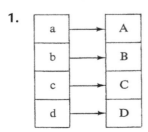

2.

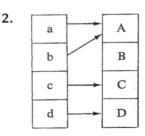

3.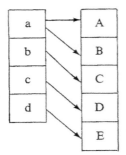

4.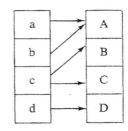

5.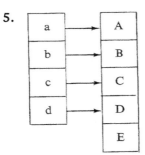

6.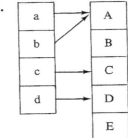

7.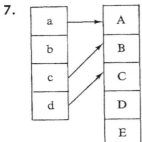

8.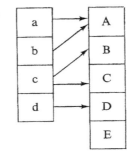

9.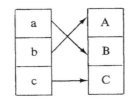

10.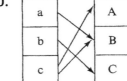

11.

12.

x	1	2	3	4
y	6	7	8	9

13.

x	1	2	3	4
y	6	6	7	8

14.

x	1	1	3	4
y	6	7	8	9

15. Determine whether or not the following assignments constitute a function.

    a. The assignment between students in a class and their heights.

    b. The assignment between students in a class and their names.

    c. The assignment between names and students in a class.

    d. The assignment between stocks listed on the New York Stock Exchange and their closing prices.

    e. The assignment between closing stock prices and stocks listed on the New York Stock Exchange.

    f. The assignment between all the cars in the United States and their colors.

    g. The assignment between the prime interest lending rate and the banks located in San Francisco.

16. The equation $y = 5x + 10$ is a rule relating the variables x and y. For this equation which variable, by convention, is taken to be a member of the first set (the domain) and which variable, by convention, is considered to be a member of the second set (the range)?

3.2 MATHEMATICAL FUNCTIONS

We know that a function consists of three components: a domain, a range, and a rule. The domain and range can be any two sets (people, cars, colors, numbers, etc.), while the rule can be given in a variety of ways (arrows, tables, words, etc.). In business situations, the primary concern is with sets of numbers (representing price, demand, advertising expenditures, cost, or profit, etc.) and rules defined by mathematical equations.

At first glance, it may seem strange to think of an equation as a rule, but it is. Consider two identical sets of real numbers and the equation $y = 15x + 10$, where x represents a number in the domain and y represents a number in the range. The equation is nothing more than the rule "Multiply each element in the domain by 15 and add 10 to the result." Similarly, the equation $y = x^2 - 7$ is the rule "Square each element in the domain and then subtract 7 from the result."

Whenever we have two sets of numbers and a rule given by an equation, where the variable x denotes an element in the domain and the variable y denotes an element in the range, we simply say that y *is a function of* x and write $y = f(x)$, although symbols other than x and y labels are frequently used when they are more appropriate to a particular problem.

Notation

The equation $y = f(x)$ states that an input x is transformed into an output y according to a rule f. The notation $f(x)$, is read as "f of x" to signify that a rule f operates on an input value x, and is used most often when the assignment rule is an algebraic formula. For example, the formula $f(x) = x^2$ is shorthand for the rule "Assign to each number its square." Similarly, the formula $f(x) = 2x - 5$ is shorthand for the rule "Multiply each value by 2 and then subtract 5 from the result." Mathematically, the value in the range associated with a particular value of x is found by substituting the value for x into the formula. Thus, $f(3)$ is the effect of applying the assignment rule to the input value 3, while $f(5)$ is the effect of applying the assignment rule to the input value 5. Effectively, the variable x is only a placeholder.

When the elements in the domain are denoted by some other variable, say P, and the elements in the range are denoted, say by D, we would write $D = f(P)$, read "D is a function of P." Care must be taken, however, not to interpret the notation $y = f(x)$ as "y equals f times x." Simply put, $y = f(x)$ is a shorthand notation that says y depends on x in a manner that satisfies Definition 3.1; that is, it is a stand-in for the rather cumbersome statement, "We have two sets of numbers and a rule which satisfies Definition 3.1; the rule is given by a mathematical equation where x and y denote elements in the domain and range, respectively." If no domain or range is given, they are assumed to be, by default, the set of all real numbers, respectively.

Example 1 Find $f(2), f(0),$ and $f(-1)$ if $f(x) = 2x^2 - 3x + 4$.

Solution This function assigns to each value of x the number obtained by squaring x and multiplying by 2, subtracting 3 times x from this result, and then adding 4. For values of this function when $x = 2$, $x = 0$, and $x = -1$, we have: then

$$f(2) = 2(2)^2 - 3(2) + 4 = 8 - 6 + 4 = 6$$

$$f(0) = 2(0)^2 - 3(0) + 4 = 0 - 0 + 4 = 4$$

and

$$f(-1) = 2(-1)^2 - 3(-1) + 4 = 2 + 3 + 4 = 9$$

Because x is only a placeholder, it can be replaced by any other quantity, as long as we are consistent in replacing every x by the new quantity. For example, if the function $f(x) = 8x$ is to be evaluated at $x =$ hot-dogs, then $f(\text{hot-dogs}) = 8\text{hot-dogs}$. Similarly for $x = \Delta x$, where Δx is simply a combination of the characters Δ and x, $f(\Delta x) = 8\Delta x$. Also $f(\$@q) = 8\$@q$. Although these last examples are contrived, they serve to indicate the placeholder quality of the variable x.

Example 2 Find $f(-2), f(z), f(h + 1)$, and $f(\Delta x)$ if $f(p) = p^3 - 2p$.

Solution This assignment rule assigns each value of p to the value obtained by cubing p and subtracting 2 times the value of p. In particular,

$$f(-2) = (-2)^3 - 2(-2) = -8 + 4 = -4$$

$$f(z) = z^3 - 2z$$

$$f(h + 1) = (h + 1)^3 - 2(h + 1)$$
$$= (h^3 + 3h^2 + 3h + 1) - (2h + 2)$$
$$= h^3 + 3h^2 + h - 1$$

and

$$f(\Delta x) = (\Delta x)^3 - 2\Delta x.$$

The symbols chosen to represent the input of a function and those selected for the output can be any letter, symbol, or acronym that makes sense for the application. Such labels are called *variables*. The input variables are further referred to as *independent variables*. This term emphasizes the fact that the values for these variables can be chosen independently of one another from the domain. The output variable for the function is then called the *dependent variable* because its value depends completely on the choice made for the input variable and the assignment rule.

One cannot always tell by looking at an equation which variable is the independent variable and which is the dependent variable. Because the independent variable comes from the domain, while the dependent variable represents an element in the range, the decision is equivalent to determining which set is the domain and which is the range. As mentioned in Section 3.1, this choice is dictated by the physical situation that the equation models.

As an example, consider a hypothetical equation that relates coal production to steel production. For the coal company, the amount of coal mined depends on the demand from the steel manufacturers who use the coal to make steel.

Once the decision has been made as to which quantity is the independent variable and which is the dependent variable, and again we stress that this choice is dictated by the physical situation, we can use our notation to communicate this decision to others. Conventionally, the equation $y = f(x)$ indicates that x is the independent variable and y is the dependent variable. Similarly, $D = f(P)$ indicates that P is the independent variable and D is the dependent variable. Note that $D = f(P)$ can also *be* read as "D is a function of (depends on) P." However, because it is a function, D depends on P in the one-to-one relationship that is required by a function; for every value of P there is one and only one value of D.

To this point, we have said very little about the domain and the range other than that they must exist, and for mathematical functions, they must be sets of numbers. In many business situations, this is not restrictive enough.

The Domain

Consider the following demand-price equation for oranges applicable to a particular grocery chain, where P is the price of an orange in dollars:

$$D = 100\, P^2 - 340P + 289 \qquad\qquad (\text{Eq. 3.1})$$

This is a demand-price equation, plotted in Figure 3.6, that relates the price of a product, denoted here as P, in dollars, to the demand D. As illustrated in the graph, the demand for oranges decreases as the price increases from $0 to approximately $1.70. Beyond $1.70, however, a strange relationship occurs: The demand for oranges increases with an increase in price. If the chain owner blindly uses Equation 3.1, he or she will come to the ridiculous conclusion that a price of $1.50 per orange will result in a demand for 4 oranges and a price of $10.00 per orange will result in a demand for 6,889 oranges.

Demand vs Price

$D = 100\,P^2 - 340P + 289$

FIGURE 3.6 A suitable domain is required.

Obviously, something is wrong. Whereas Equation 3.1 represents a valid relationship between price and demand for certain prices, specifically prices between 0 and $1.70, it is not a reasonable relationship for other prices, such as $10 per orange. In representing this relationship without some indication of the values for P for which it is valid, we have neglected important information. A complete description is

$$D = 100P^2 - 340P + 289$$
(P an integer lying between 0 and 1.70 inclusive).   (Eq. 3.2)

The restriction, "P an integer lying between 0 and 1.70, inclusive," is a restriction on the domain. That is, only integers between 0 and 1.70 are considered as the domain.

The domain, therefore, is nothing more than the set of all *allowable* values from which we can select the independent variable. If the domain is not given explicitly, it is taken to be all real numbers. One exception occurs when it is clear from the given equation that certain numbers cannot be values of the independent variable.

Example 3 Determine the domain for the function $y = 1/(x - 2)$.

Solution An admissible domain for x is all real numbers except 2 because the function is not defined at $x = 2$. A complete description for this function is

$$y = \frac{1}{x - 2} \quad (x \text{ any real number except 2}).$$

The Range

For mathematical functions, we usually define the range to be the set of all numbers between minus infinity and plus infinity. Typically, not all these values are used by a given equation, but Definition 3.1 does not stipulate that every value in the range must be assigned. In Example 1 in Section 3.1 (see Figure 3.2), the value 4 was not used. Note that we can always pinpoint the values of the dependent variable used by applying the assignment rule to all the elements in the domain.

Example 4 For the function given by

$$y = f(x) = 30x^2 + 20x + 10,\ 0 \leq x \leq 10$$

determine   **a.**   the independent variable,

             b.   the dependent variable,

             c.   the domain, and

             d.   the range.

Solution

a. x is the independent variable.

b. y is the dependent variable.

c.   The domain of the function consists of all real numbers between 0 and 10.

d.   The range of the function consists of all real numbers between plus and minus infinity. Note, however, that only y-values lying between 10 and 3,210 are actually used by this function.

The Vertical Line Test

A graph that displays the domain, range, and the relationship between these two sets of values can easily be used to determine if these three elements constitute a function. To represent a function, a graph must pass the *vertical line test*.

This test requires that any and all vertical lines that cross the x-axis within the domain must intersect the graph of the function at one and only one point. This ensures that each value of x in the domain corresponds to exactly one element in the range.

The graph shown in Figure 3.6 passes this test because each vertical line through the x-axis intersects the graph at one and only one point. The graph is shown in Figure 3.7, however, does not pass the vertical line test because there is at least one vertical line, shown as dotted, that crosses the graph at more than one y value.

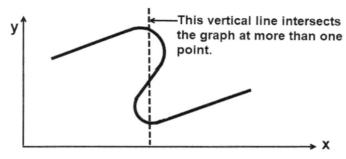

FIGURE 3.7 The vertical line test.

Exercises 3.2

1. Consider the function $D(z) = z^2 - 30z + 225$, $0 \leq z \leq 10$, z an integer.

   a. What is the domain of the function? (List the notation he values.)

   b. What is the range of the function?

   c. What values of the range are actually taken by D? (List them.)

2. The following equations relate values of x to values of y, for each equation, list a possible domain and range to qualify the sets of numbers and the equation as a function.

   a. $y = 3 + 4x$ **b.** $y = 1/(x - 3)$

   c. $y = 10 + 18x^2$ **d.** $y = x^2 - 1/x$

   e. $y = (x + 4)/(x + 2)$ **f.** $f(x) = (x + 2)/[(x + 3)(x - 6)]$

   g. $f(x) = x^2 + 3x + 10$ **h.** $f(x) = (x^2 + 3)/(x + 5)$

3. Determine whether or not the relationship defined by $y = 2 + 3x$, $0 \leq x < \infty$ is a function. Determine whether the inverse relationship is a function. (*Hint:* Solve for x in terms of y.)

4. **a.** Determine whether or not the relationship $y = +\sqrt{x}$ is a function for $0 \leq x < \infty$.

   b. Determine whether or not the relationship $y = \pm\sqrt{x}$ is a function for $0 \leq x < \infty$.

5. Determine whether or not $z = 2w^2 + 4$, $0 \le w \le 4$, is a function. Is the inverse relationship a function for $4 \le z \le 10$? (Hint: solve for w in terms of z.)

6. Given the function $f(x) = x^2 + 2x + 3$, find

   a. $f(2)$ **b.** $f(0)$ **c.** $f(10)$ **d.** $f(-1)$ **e.** $f(1/x)$

7. Given the function $y = f(x) = x^2 + 3x - 6$, find

   a. $f(2)$ **b.** $f(5)$ **c.** $f(0)$ **d.** $f(a+b)$ **e.** $f(x + \Delta x)$

8. Given the function $f(x) = x^3 + 6x - 4$, find

   a. $f(0)$ **b.** $f(1)$ **c.** $f(3)$ **d.** $f(5)$ **e.** $f(a+b)$

   f. $f(x^2)$ **g.** $f(x + \Delta x)$

9. Given the function $f(a) = a + 2a^2 + a^3$, find

   a. $f(2)$ **b.** $f(d)$ **c.** $f(x + y)$ **d.** $f(2a)$

10. A store owner has determined that the demand for a particular brand of specialty shoes is related to price by the function $D = p^2 - 28p + 196$. Determine a domain for this equation so that the resulting function represents a plausible demand curve.

3.3 POLYNOMIAL FUNCTIONS

Mathematical functions are the core of real-world applications and essential tools for decision makers. In this section, we present a class of one of the most useful business functions – polynomials

A function $f(x)$ is a *polynomial function* if it has the form

$$f(x) = a_n x^n + a_{n-1} x^{n-1} + \cdots + a_2 x^2 + a_1 x + a_0 \qquad \text{(Eq. 3.3)}$$

here a_n, a_{n-1}, $\cdots$, a_2, a_1, and a_0 are all known numbers.

The powers of x (*that is, n, n − 1*, and so on) are required to be nonnegative integers[1], and the highest power with a corresponding non-zero coefficient a, is called the *degree* of the polynomial function. The lead coefficient, a_n, is the coefficient of the x term with the highest power and cannot be zero, but any of the other following coefficients can be. The constant term a_0 is the coefficient of $x^0 = 1$.

---

[1] A function where n is a negative integer is defined as a *rational function*. The case where n is either positive or negative, but not an integer is defined as a *power function*. These two function types are presented at the end of this section.

Example 1 Determine which of the following functions are polynomial functions. For those that are, state their degree, and coefficients.

a. $f(x) = 3x^2 - 2x$ **b.** $f(x) = 0.8x^5 - 2.25x^3 - \sqrt{7}$

c. $f(x) = \sqrt{x}$ **d.** $f(x) = \dfrac{1}{3}$ **e.** $f(x) = \dfrac{1}{x}$

Solution

a. This is a polynomial function of degree 2, with $a_2 = 3$, $a_1 = -2$, and $a_0 = 0$.

b. This is a polynomial function of degree 5, with $a_5 = 0.8$, $a_4 = 0$, $a_3 = 2.25$. $a_2 = a_1 = 0$, and $a_0 = \sqrt{7}$.

c. This is *not* a polynomial function because $\sqrt{x} = x^{\frac{1}{2}}$. Here x is raised to the $\dfrac{1}{2}$ power, which is not a non-zero integer.

d. This is a polynomial function of degree 0, with $a_0 = \dfrac{1}{3}$.

e. This is *not* a polynomial function because $\dfrac{1}{x} = x^{-1}$. Here x is raised to the -1 power, which is a *negative* integer (this is a rational function, as presented at the end of this section).

Constant Functions

A zero-degree polynomial, which has the form $f(x) = a$, is referred to as *constant function*. These functions are more commonly written as

$$y = c \qquad \text{(Eq. 3.4)}$$

A constant function says that every input value of x is assigned the same non-zero constant c. We know from Section 2.3 that the graph of an equation of the form $y = c$ is a horizontal straight line. Thus, as all such equations are zero-degree polynomial functions, it follows that the graphs of all zero-degree polynomials are horizontal straight lines. Two such straight-line graphs are shown in Figure 3.8.

It should be noted that the *x-axis* in a Cartesian coordinate system is the graph of the *zero function* $y = 0$. However, this particular function *is not* a polynomial function because the condition that the highest coefficient, in this case, a_0, cannot be 0 is violated. This function has no degree because every power of the variable x is multiplied by a zero coefficient.

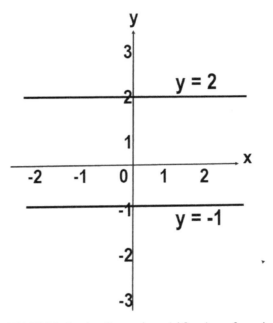

FIGURE 3.8 Graphs of two polynomial functions of zero degree.

Additional Linear Functions

A first-degree polynomial is a function of the form $f(x) = a_1 x + a_0$, with $a_1 \neq 0$. The constants a_1 and a_0 are typically replaced by the symbols m and b, respectively, and the function is more typically written as:

$$y = mx + b \quad (m \neq 0) \qquad \text{(Eq. 3.5)}$$

This last equation is the slope-intercept form of a straight line that was presented in Section 2.3. The graph of this equation is a straight line with slope m; it intercepts the y-axis at the point $(0, b)$. Thus, the graph of a first-degree polynomial is a straight line with a non-zero slope. The graph of the line $y = -100x + 1500$ is shown in Figure 3.9. The line intercepts the y-axis at $(0, 1500)$ and has a slope $m = -100$.

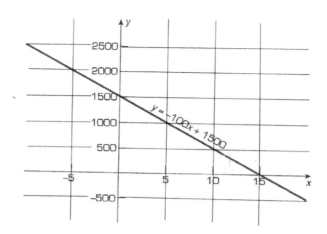

FIGURE 3.9 The graph of the first-degree polynomial $y = -100x + 1500$.

Functions whose graphs are nonvertical straight lines – the zero function, polynomials of degree zero, and polynomial functions of degree 1 – are all *linear functions*. By far, some of the most important business applications are modeled using linear equation, and were examined extensively in Chapter 2.

A vertical line, such as the one shown in Figure 3.10, has an undefined slope and, as such, *is not* a linear function.

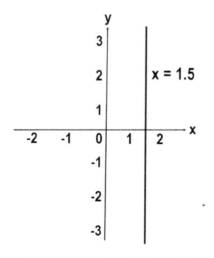

FIGURE 3.10 The vertical line $x = 1.5$.

Rational and Power Functions

One class of function related to polynomial functions are **rational functions**. A rational function is one that is the quotient of two polynomials. It has the form:

$$f(x) = \frac{p(x)}{q(x)} \qquad \text{(Eq. 3.6)}$$

where $p(x)$ and $q(x)$ are both polynomial functions

Example 2 Determine which of the following functions are rational functions.

a. $f(x) = \dfrac{8x^5 - 4x^3 + 2}{x^2 - 1}$ 　　　　　　 **b.** $f(x) = \dfrac{x^2 - 3}{x^3 + 6}$

c. $f(x) = \dfrac{\sqrt{x} + 3}{2x}$ 　　　　　　 **d.** $f(x) = \dfrac{1}{x}$

Solution

a. This is a rational function. Here, the numerator is the 5th degree polynomial function $p(x) = 8x^5 - 4x^3 + 2$, and the denominator is the 2nd degree polynomial $q(x) = x^2 - 1$.

b. This is a rational function because the numerator and denominator are polynomial functions of degrees 2 and 3, respectively.

c. This *is not* a rational function. Although the denominator is a polynomial function of degree 1, the numerator is not a polynomial function due to the $x^{1/2}$ term, which raises x to a non-integer power.

d. This is a rational function because the numerator, $p(x) = 1$, is a polynomial function of degree zero and the denominator, $q(x) = x$, is a polynomial function of degree 1.

Because a rational function is a quotient it is not defined when its denominator is zero. Thus, the largest domain of the rational function $f(x) = p(x)/q(x)$ is the set of all real numbers for which $q(x) \neq 0$.

Example 3 What values must be excluded from the domain of the rational function $f(x) = \dfrac{x^3 - 2x^2 + 5}{x^2 - 4}$.

Solution This function is the quotient of two polynomials, which qualifies it as a rational function. However, the denominator of the function is zero when $x = \pm 2$. Thus, these two values must be excluded from the domain.

Example 4 What is the largest domain possible for the rational function

$$f(x) = \frac{1}{x}$$

Solution The denominator of this rational function is zero when $x = 0$. Thus, the largest domain is the set of all non-zero real numbers.

A second class of functions related to both polynomial and rational functions are **power functions**. A power function has the form

$$f(x) = ax^p \qquad \text{(Eq. 3.7)}$$

where a is any non-zero value and p is a known real number.

When p is an integer, positive or negative, the power function becomes a polynomial or rational function, respectively. Thus, if p is zero or a positive integer n, Equation 3.7 reduces to Equation 3.3, and $f(x) = ax^n$, a polynomial function of degree n. Conversely, when p is a negative integer, $-n$, Equation 3.7 reduces to $f(x) = x^{-n} = 1/x^n$, which is a rational function.

Exercises 3.3

In Exercises 1 through 18, determine which relationships are polynomial functions and, for those that are, state their degree.

1. $f(x) = 2.1x^2$

2. $f(x) = 2x^{2.1}$

3. $f(x) = 3x^2 - 2.1$

4. $f(x) = 2x^3 - 2.1$

5. $f(x) = \pi$

6. $f(x) = \pi^3 - 3$

7. $f(x) = \pi^{2.1}$

8. $f(x) = x^\pi$

9. $f(x) = 1 - x^5$

10. $f(x) = x^3 - 3x^4$

11. $f(x) = \dfrac{3x^4 + 2x^2 - 0.5}{2 - x}$

12. $f(x) = \dfrac{3x^4 + 2x^2 - 0.5}{2 - \sqrt{x}}$

13. $f(x) = \dfrac{3x^4 + 2x^2 - 0.5}{\sqrt{2} - x}$

14. $f(x) = \dfrac{3x^4 + 2x^2 - 0.5}{\sqrt{2} - \sqrt{3}}$

15. $f(x) = \dfrac{3x^4 + 2x^2 - 0.5}{\sqrt{2 - x}}$

16. $f(x) = \dfrac{3x^4 + 2x^2 - 0.5}{2 - x^2}$

17. $f(x) = \dfrac{x^2 + 2x - 3}{x^2 + 3x + 5}$ **18.** $f(x) = \dfrac{x^4 + x^2 + 5}{x^6 - 3x^7 - 5x^8}$

19. Determine which of the functions in exercises 1 through 18 are rational functions.

20. Determine which of the functions in exercises 1 through 18 are power functions.

21. What is the degree of the product of two polynomial function?

22. Show, by example, that it is possible to add two polynomials functions of degree 2 and obtain a polynomial of degree 1.

23. Prove that every polynomial function must intersect the y-axis.

24. The relationship between, F, the temperature in degrees Fahrenheit, and its equivalent temperature, C, in degrees Celsius, is given by the formula

$$C = \frac{5}{9}(F - 32)$$

   a. Is C a polynomial function of the variable F? If so, what is its degree?

   b. Determine the Celsius equivalent of $90°$ Fahrenheit.

   c. Determine the Fahrenheit equivalent of 212 degrees Fahrenheit.

   d. Solve for F in terms of C.

3.4 QUADRATIC FUNCTIONS

An important class of functions that are more complex than first degree linear functions and their resulting straight-line graphs are second-degree polynomials. These functions are referred to as *quadratic functions*, and have the form:

$$f(x) = a_2 x^2 + a_1 x + a_0$$

where $a_2 \neq 0$. If we replace the constants a_2, a_1, and a_0, by a, b, and c, respectively, this second-degree polynomial function is written in its more conventional form as:

$$y = ax^2\, bx + c \qquad\qquad (\text{Eq. } 3.8)$$

In Equation 3.8, the variable that is squared is referred to as the **quadratic variable**, which in this case is x. Note that what determines if an equation is a function are *not the symbols* used in the equation, but whether the equation, domain, and range satisfy the definition of a function provided in Section 3.1.

Example 1 Determine which of the following functions are quadratic functions. For those that are, state their coefficients, a, b, and c.

a. $y = 2x^2 - \frac{1}{2}$ **b.** $y = 3x - x^2$ **c.** $n^2 = 2p + 4$

Solution

a. This is a quadratic function in the variable x with $a = 2$, $b = 0$, and $c = -1/2$.

b. Rewriting this equation as $y = -x^2 + 3x$, we see that this is a quadratic function in the variable x with $a = -1$, $b = 3$, and $c = 0$.

c. Rewriting this equation as $f(p) = 1/2 \, n^2 - 2$, we see that it is a quadratic function in the variable n, with $a = \frac{1}{2}$, $b = 0$, and $c = -2$.

As in the case of linear equations and in part (c) of this example, the letters y and x used in Equation 3.8 are arbitrary; any other two letters are equally appropriate. The essential point is the form of the relationship between the variables. That is, a quadratic equation is one in which one variable can be written as the sum of a constant times the second variable squared, plus a constant times the second variable, plus a constant.

The graph of a quadratic function is a parabola, which is a shape similar to the cone of a rocket. Figures 3.11 and 3.12 are graphs of two different quadratic function.

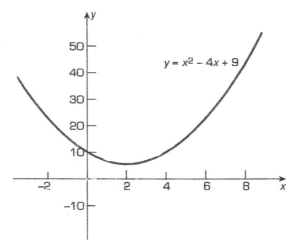

FIGURE 3.11 The graph of the quadratic function $y = x^2 - 4x + 9$.

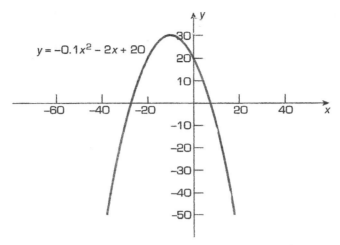

FIGURE 3.12 The graph of the quadratic function $y = -0.1x^2 - 2x + 20$.

Example 2 From past experience, a washing machine manufacturer knows that the relationship between its monthly profit, P, and the number, n, of machines produced is expressed by the equation

$$P = -22n^2 + 210n - 100$$

where P is in thousands of dollars and n is in thousands of machines. Graph this function and explain the significance of the curve.

Solution
Evaluating the function for various values of x, we obtain Table 3.2.

TABLE 3.2 Profit versus Number of Machines Manufactured and Sold.

n (in thousands)	0	1	2	3	4	5	6	7	8	9	10
P (in $ thousands)	−100	88	232	332	388	400	368	292	172	8	−200

The points listed in Table 3.2 are plotted in Figure 3.13, with the graph showing positive values of n only because negative production ($n < 0$) has no practical significance. Thus, the domain of the function is $n \geq 0$.

Production of 1,000 washing machines ($n = 1$) results in a profit of $88,000. Profits continue to increase with higher production levels until approximately $n = 5$, which corresponds to 5,000 machines being produced. At this point, the corresponding profit of $400,000 is near its maximum. Increasing production beyond this starts to reduce profits. One explanation is that the marketplace can absorb approximately 5,000 washing machines monthly from this producer[2]. Additional machine production would increase costs but add less to income. Thus, profits are reduced.

If no washing machines are produced, $n = 0$, and the corresponding profit is

$$P = -22(0)^2 + 210(0) - 100 = -\$100 \text{ thousand dollars}$$

That is, the manufacturer experiences a loss of $100,000. This loss reflects fixed costs such as rent, insurance, and leasing payments on equipment.

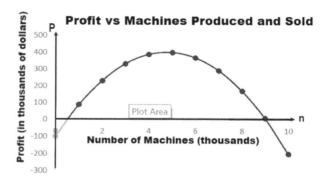

FIGURE 3.13 Graph of the function $P = -22n^2 + 210n - 100$.

Solving for the Quadratic Variable

In general, whenever we wish to solve a quadratic equation, it is easier to select values of the variable that is squared (x in Equation 3.8) and n in Example 2, and then use the given equation to find the value of the second variable, rather than the other way around, Sometimes, however, we have no choice. As an example of this, consider the following:

---

[2] The actual maximum profit is achieved when 4,770 machines are produced and sold. This value is easily determined using the optimization technique presented in Section 6.1.

Example 3 Based on observations of prices, the demand D for oranges at a local fruit stand satisfies the equation $D = -0.25P^2 + 6P + 900$, where P is the price per orange (in cents). On a given Saturday morning, the store has 100 oranges in stock. Determine the price the store should charge for oranges if it wishes to deplete its inventory by the end of the day.

Solution Here, we seek the price that *results in zero* inventory. Mathematically, this means we are asked to find the value of the quadratic term P, for a given value of 100 for the linear term, D. Substituting $D = 100$ into the demand-price equation, we find that P must satisfy the quadratic equation

$$100 = -0.25P^2 + 6P + 900,$$

which can be rewritten as

$$0.25P^2 - 6P - 800 = 0$$

Solving this requires using the quadratic equation,[3] with $a = 0.25$, $b = -6$, and $c = -800$ (see the chapter appendix if you are not familiar with the quadratic formula). Using these values in the quadratic formula we obtain:

$$P_1 = \frac{-(-6) + \sqrt{(-6)^2 - 4(.25)(-800)}}{2(0.25)} = \frac{6 + \sqrt{36 + 800}}{.5}$$

$$= \frac{6 + \sqrt{836}}{0.5} = \frac{6 + 28.91}{0.5} = \frac{34.91}{0.5} = 69.82$$

and

$$P_2 = \frac{6 - \sqrt{836}}{0.5} = \frac{6 - 28.91}{0.5} = \frac{-22.91}{0.5} = -45.82$$

As the negative solution has no practical meaning, the oranges should be priced at 69¢ a piece. This creates a demand of

$$D = -0.25(69)^2 + 6(69) + 900 = 123, \text{ which will deplete the stock.}$$

A price of 70¢ will only create a demand of

$$D = -0.25(70)^2 + 6(70) + 900 = 95, \text{ which will not deplete the stock.}$$

---

[3] We assume that most students are familiar with the quadratic formula. However, for those that are not, or need a review, the quadratic formula is presented in this chapter's appendix.

Example 4 A warehouse has 12,000 cans of a discontinued tennis ball which it wants to liquidate in a week. From past experience, it is known that demand D per week (in cans) is related to the price P (in dollars) by the equation

$$D = -3{,}000P^2 - 13{,}500P + 27{,}000.$$

Determine the price that will result in zero inventory.

Solution We seek the price P that will yield $D = 12{,}000$. Substituting this value into the given equation, we find that P must satisfy

$$12{,}000 = -3{,}000P^2 - 13{,}500P + 27{,}000$$
$$3{,}000P^2 + 13{,}500P - 15{,}000 = 0$$
$$3P^2 + 13.5P - 15 = 0.$$

Using the quadratic formula with $a = 3$, $b = 13.5$, and $c = -15$, we obtain

$$P_1 = \frac{-13.5 + \sqrt{(13.5)^2 - 4(3)(-15)}}{2(3)} = \frac{-13.5 + \sqrt{182.25 + 180}}{6}$$

$$= \frac{-13.5 + \sqrt{362.25}}{6} = \frac{-13.5 + 19.03}{6} = \frac{5.53}{6} = 0.922$$

and

$$P_2 = \frac{-13.5 - \sqrt{362.25}}{6} = \frac{-13.5 - 19.03}{6} = -5.42$$

Again, the negative price has no practical meaning, and the tennis balls should be wholesaled at 92¢ per can.

Exercises 3.4

1. Determine which of the following equations are quadratic, and for those that are, specify the quadratic variable.

   a. $x^2 - x = y$

   b. $y^4 = 3$

   c. $x^2 - 2x + 2 = y$

   d. $y - x^2 = 0$

   e. $y + x = 3$

   f. $n^2 = 2d + 5$

   g. $R = 2S^2$

   h. $\sqrt{y} = x$

   i. $y = x^2 + 2x$

   j. $x = y^2$

   k. $x^2 - 6x + 9 = y$

   l. $x - 2y = 4$

2. Graph the following quadratic curves by first plotting a sufficient number of points to determine the curve's correct shape.

   a. $y = x^2$

   b. $y = -x^2$

   c. $y = -x^2 - x + 4$

   d. $y = x^2 + x - 4$

   e. $d = 2p^2 - 8$

   f. $y = x^2 + 3$

   g. $y = 1 - x^2$

   h. $y = x^2 - 3x + 2$

3. Use the quadratic formula to find the values of x that satisfy the following equations:

   a. $x^2 - x - 6 = 0$

   b. $3x^2 - 2x - 5 = 0.$

   c. $4x^2 - 7 = 0$

   d. $1/3x^2 - x - 1 = 0$

   e. $x^2 - x = 4$

   f. $3x^2 - 12x + 6 = 0$

   g. $x^2 - x - 2 = 0$

   h. $x^2 - x = 0$

   i. $3x^2 - 12x = 0$

4. Ogden Motors has 15 identical model automobiles which it wants to sell within a month. From past experience, it is known that the demand, d, per month is related to the price p (in dollars) by the equation $d = 0.04p^2 + 10{,}000$. Determine the maximum price that will result in no inventory at the end of a month.

5. Ogden Motors uses the formula $V = (-0.005t^2 - 0.05t + 0.8)P$ to compute the value of used cars, where V denotes the current used car price (in dollars), t denotes the age of the car (in years), and P denotes the price when the car was new (in dollars).

   a. Find the value of a $25,000 automobile after 3 years.

   b. Find the value of a $25,000 automobile immediately after it leaves the showroom.

   c. Determine the time when a $40,000 automobile will be worth $12,800.

6. A manufacturer of automatic fire alarm systems determines that its total cost is given by $C = 1{,}000x^2 + 5{,}000x + 10{,}000$. Each system sells for $12,000. Determine the break-even point for this process both. graphically, and b. algebraically.

7. A small company that manufactures reusable 12-ounce plastic cups has determined that its monthly profit (in millions of dollars) is given by $P(x) = -x^2 + 30x - 60$, where x is the number of 12-ounces cup sold (in millions) per month.

   a. Graph the profit equation.

b. Determine the largest number of cups that the company can sell and still make a profit?

c. If the company sells more than the number of cups in part (b), explain how it is possible for it to lose money.

d. Determine the maximum profit.

8. A manufacturer has determined that the yearly profit P (in dollars) is directly related to the number of units sold n by the formula $P = n^2 - 400n - 50,000$.

a. Graph this equation by plotting the points corresponding to $n = 0$, 100, 200, 500, 600, 900, and 1000.

b. Determine the loss if no units are sold.

c. Determine the profit if 1,000 units are sold.

d. How many units must be sold if a profit of $100,000 is desired?

3.5 EXPONENTIAL FUNCTIONS

Straight-line and quadratic functions are some of the simplest and yet valuable function in business and science. By themselves, however, they are not sufficient for modeling all real-world phenomena. Many such processes follow other functions. One of the most important of these remaining functions is the exponential function, which is a keystone of modern portfolio theory and environmental science.

In particular, most natural phenomena can be accurately modeled or represented by an exponential function. Examples of such situations are pollution levels, the use of natural resources, and the radioactive decay of certain materials. In practice, phenomena such as these can be misleading because their graphs stay relatively constant or flat for many years, very much like the graph of a linear equation. As the value of the exponent builds, however, the value of the y variable suddenly "takes off" beyond any expectation based on what a linear or quadratic model would predict. Such a situation is shown in Figure 3.14, which illustrates the pollution level of nitrogen oxide versus time (in centuries).

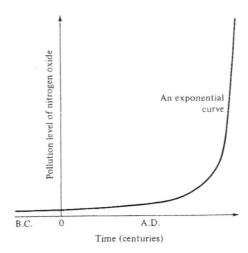

FIGURE 3.14 A typical exponential curve.

An *exponential function* in the variable x is a function having the form

$$f(x) = a\ (b^x)\ x \text{ a real number}$$

and is typically written using the form as

$$y = a(b^x) \tag{Eq. 3.9}$$

where a is a known non-zero real numbers and b is a positive real number not equal to 1. The number b is called the *base*. The distinguishing feature of an exponential function and the reason for its name is that the variable x is the exponent

Example 1 Determine which of the following are exponential functions. For those that are, give the values of the base b and the constant a.

a. $f(x) = 5^x$ **b.** $f(x) = 2.1(5^x)$ **c.** $y = \left(\dfrac{1}{3}\right)^x$

d. $f(x) = -2\left(\dfrac{1}{3}\right)^x$ **e.** $f(x) = (-2)^x$ **f.** $y = x^5$

Solution
a. This is an exponential function with the base $b = 5$ and the constant $a = 1$.
b. This is an exponential function with the base $b = 5$ and the constant $a = 2.1$.

c. This is an exponential function with the base $b = 1/3$ and the constant $a = 1$.

d. This is an exponential function with the base $b = 1/3$ and the constant $a = -2$.

e. This is *not* an exponential function because the base $b = -2$ is not a positive number.

f. This is *not* an exponential function but rather a polynomial function of degree 5, as well as a power function.

The base b is not allowed to be 1 in Equation 3.9 because if $b = 1$, then

$$f(x) = a(b)^x = a(1)^x = a(1) = a$$

which is a polynomial function of degree zero.

Exponential functions with $a > 0$ will always have one of the two general shapes shown in Figures 3.15 and 3.16. Such graphs never intersect the x-axis but approach it from above in one direction and increase steeply in the other. These graphs always intersect the y-axis at the point $(0, a)$.

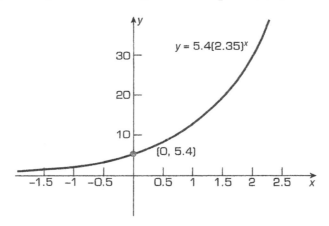

FIGURE 3.15 An exponentially increasing curve.

Figure 3.15 is the graph of the function $f(x) = 5.4(2.35)^x$. Here the base $b = 2.35$, which is greater than 1, and the coefficient a is positive and equal to 5.4. This graph is said to *increase exponentially*. If we restrict our attention to the first quadrant, the value of the function becomes very large very quickly as the value of x increases.

Figure 3.16 illustrates the shape of an exponential curve when the base is between 0 and 1 while a is still a positive value. This is the graph of the function $y = 2(1/3)^x$. This function is said to *decrease exponentially*.

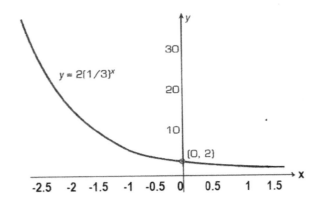

FIGURE 3.16 An exponentially decreasing curve.

Just as polynomial functions are characterized by their degree, exponential functions are characterized by their base. Similarly, as polynomial functions of degrees 0, 1, and 2 are the most prevalent for modeling whene polynomial functions are required, some bases are more valuable when exponential functions are used. The most useful of these bases is the irrational number

$$b = 2.718281828459\cdots$$

known as Euler's number (pronounced "Oiler"), which is named after the mathematician Leonhard Euler. Its graph is shown in Figure 3.17.

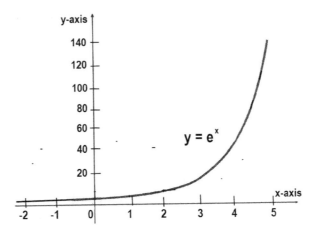

FIGURE 3.17 The graph of the exponential function $f(x) = e^x$.

Example 2 The number of a certain type of bacteria placed in a culture dish at room temperature grows as given by the equation:

$$Number\ of\ Bacteria = (Original\ Number\ of\ Bacteria)\ e^{0.2t}$$

where t is the time, in hours that the culture has been in the room, and e, Euler's number, is the irrational number $b = 2.7182818\cdots$.

Using this equation, determine the number of bacteria that will be in this culture after **a.** 10 hours and after **b.** 2 days

Solution

a. *Number of bacteria after 10 hours*

$$= 100(2.71818)^{.2(10)} = 100(7.3885) = 738.85$$

b. *Number of bacteria after 2 days*

$$= 100(2.71818)^{.2(48)} = 100(14,759.4727) = 1,475,947.27$$

Notice the number of bacteria has grown to 738 after ten hours but has increased to close to one and one-half million after 2 days.

Exercises 3.5

In Exercises 1 through 12, determine whether the function is (or can be written as) an exponential function. For those that are, list the values of a and b:

1. $f(x) = 2(7)^x$ **2.** $f(x) = 7(2^x)$ **3.** $f(x) = 2x^7$

4. $f(x) = 7^{2x}$ **5.** $f(x) = (7x)^2$ **6.** $f(x) = 2\pi^x(x)$

7. $f(x) = \pi e^x$ **8.** $f(x) = e^\pi$ **9.** $f(x) = 7^x$

10. A model of worldwide population growth, in billions of people, since 2010 is given by this formula, where e known as Euler's (pronounced *Oiler's*) number," is the irrational number $b = 2.7182818\cdots$.

$$Population = 7.5\ e^{0.02(Year - 2010)}$$

Using this formula, estimate the worldwide population in

a. 2025

b. 2030

11. The number of a certain type of bacteria placed in a culture dish at room temperature grows as is given by the equation:

    Number of Bacteria = (Original Number of Bacteria) $e^{0.15t}$

    where t is the time, in hours that the culture has been in the room, and e, known as Euler's (pronounced *Oiler's*) number, is the irrational number $b = b = 2.7182818\cdots$. Using this equation, determine the number of bacteria in this culture for:

    a.   an original number of 200 bacteria after 15 hours

    b.   an original number of 200 bacteria after 30 hours

12. The number of remaining bacteria in a certain culture that is subject to refrigeration can be approximated by the equation:

    Remaining Bacteria = (Original Number of Bacteria) $e^{-0.032t}$

    where t is the time, in hours, that the culture has been refrigerated, and e known as Euler's (pronounced *Oiler's*) number, is the irrational number $b = 2.7182818\cdots$. Using this equation, determine the number of remaining bacteria in this culture for:

    a.   an original number of bacteria of 300,000 that is refrigerated for 10 hours

    b.   an original number of bacteria of 300,000 that is refrigerated for 24 hours

    c.   an original number of bacteria of 500,000 that is refrigerated for 48 hours

    d.   an original number of bacteria of 500,000 that is refrigerated for 72 hours

13. A model to estimate the number of grams of a radioactive isotope left after t years is given by the formula:

    Remaining Material = (*Original Material*)$e^{-0.00012t}$

    where e, known as Euler's (pronounced *Oiler's*) number, is the irrational number $b = 2.7182818\cdots$. Using this formula determine the amount of radioactive material remaining after:

    a.   1000 years, assuming an initial amount of 100 grams

    b.   500 years, assuming an initial amount of 250 grams.

3.6 SUMMARY OF KEY POINTS

Key Terms

- Degree of a polynomial
- Domain
- Euler's number, e
- Exponential function
- First-degree polynomial function
- Function
- Linear function
- Polynomial function
- Power function
- Quadratic formula
- Quadratic function
- Rational function
- Second-degree polynomial function
- Zero function

Key Concepts

Functions

- A function consists of two sets and an assignment rule between them, which assigns every value in the first set a unique, but not necessarily different, element of the second set.
- Functions can be specified by words, equation, graphs, or tables.
- When a graph depicts a function, the domain is always placed on the horizontal axis and the range on the vertical axis. The assignment rule assigns a number on the vertical axis to each value on the horizontal axis.
- A graph represents a function if and only if the graph passes the vertical line test. This test requires that any and all vertical lines that cross the horizontal axis at a value in the domain must intersect the graph at one and only one point.

Polynomial, Rational, and Power Functions

- The graph of a polynomial function of degree zero is a horizontal line.
- The graph of a polynomial function of degree one is a straight line with a non-zero slope.

- A quadratic function is a polynomial of degree two. Its graph is a parabola.
- Unless stipulations are made to the contrary, the domain of a rational function is the set of all real numbers at which the function's denominator is non-zero.
- A power function has the form $f(x) = ax^p$ where a is any non-zero value and p is a known real number. When p is an integer, positive or negative, the power function becomes a polynomial or rational function, respectively.

Exponential Functions

- The graph of an exponential function of the form $y = a(b^x)$ approaches the horizontal axis from above when $a > 0$; it increases steeply when $b > 1$ and decreases steeply when $0 < b < 1$. When $a < 0$ the exponential function $y = a(b^x)$ approached the horizontal axis from below; it decreases steeply when $b > 1$ and increase steeply when $0 < b < 1$.

3.7   CHAPTER APPENDIX: THE QUADRATIC FORMULA

The solutions to equations of the form $ax^2 + bx + c = 0$ are given by the *quadratic formula*[4]

$$x = \frac{-b \pm \sqrt{b^2 - 4ac}}{2a} \qquad \text{(Eq. 3.10)}$$

To solve any quadratic equation, substitute the values of its coefficients a, b, and c into the quadratic formula and simplify.

Example 1 Solve the equation $x^2 + 2x - 3 = 0$ for x.

Solution This is a quadratic equation with $a = 1$, $b = 2$, and $c = -3$. Substituting these values into the quadratic formula, we obtain

$$x = \frac{-2 \pm \sqrt{2^2 - 4(1)(-3)}}{2(1)} = \frac{-2 \pm \sqrt{4 + 12}}{2} = \frac{-2 \pm \sqrt{16}}{2} = \frac{-2 \pm 4}{2}$$

Using the plus sign, we obtain one solution as $x = (-2 + 4)/2 = 1$. Using the minus sign, we find a second solution as $x = (-2 - 4)/2 = -3$.

---

[4] Proof of this formula is provided at the end of this section.

Example 2 Solve the equation $4y^2 - 2y = 3$ for y.

Solution We first rewrite this equation in the form as $4y^2 - 2y - 3 = 0$, which is a quadratic equation with $a = 4$, $b = -2$, and $c = -3$. Substituting these values into the quadratic formula, we have

$$y = \frac{-(-2) \pm \sqrt{(-2)^2 - 4(4)(-3)}}{2(4)} = \frac{2 \pm \sqrt{4 + 48}}{8} = \frac{-2 \pm \sqrt{52}}{8} = \frac{2 \pm 7.21}{8}$$

The solutions are then $y = (2 + 7.21)/8 = 1.15$ and $y = (2 - 7.21)/8 = -0.65$, with all calculations rounded to two decimals.

The quadratic formula does not always yield two solutions. If $b^2 - 4ac = 0$, the formula reduces to

$$x = \frac{-b \pm \sqrt{0}}{2a} = -\frac{b}{2a}$$

In these cases, the quadratic equation has only one solution. If $b^2 - 4ac$ is negative, the square root cannot be taken, and no real solutions exist. Readers familiar with complex numbers will note that this case has complex solutions. Because complex numbers have no use in commercial situations, we do not consider them here.

Example 3 Solve the equation $x^2 - 2x + 1 = 0$ for x.

Solution Here $a = 1$, $b = -2$, and $c = 1$. Substituting these values into the quadratic formula, we obtain

$$x = \frac{-(-2) \pm \sqrt{(-2)^2 - 4(1)(1)}}{2(1)} = \frac{2 \pm \sqrt{4 - 4}}{2} = \frac{2 \pm 0}{2} = 1.$$

The only solution is $x = 1$.

Example 4 Solve the equation $2p^2 + p + 1 = 0$ for p.

Solution Substituting $a = 2$, $b = 1$, and $c =$ into the quadratic formula, yields

$$p = \frac{-1 \pm \sqrt{(1)^2 - 4(2)(1)}}{2(2)} = \frac{-1 \pm \sqrt{1 - 8}}{4} = \frac{-1 \pm \sqrt{-7}}{4}.$$

Because $\sqrt{-7}$ is not defined as a real number, the given equation has no real solutions.

Proof of the Quadratic Formula

To prove the quadratic formula, we first rewrite $ax^2 + bx + c = 0$ as $ax^2 + bx = -c$, and then divide both sides of the equation by a, obtaining

$$x^2 + \frac{b}{a}x = -\frac{c}{a}.$$

Using a method known as completing the squares, we add the quantity $(b^2/4a^2)$ to both sides of this equation, which yields

$$x^2 + \frac{bx}{a} + \frac{b^2}{4a^2} = -\frac{c}{a} + \frac{b^2}{4a^2}$$

Rewriting the left side of this equation as

$$x^2 + \frac{bx}{a} + \frac{b^2}{4a^2} = \left(x + \frac{b}{2a}\right)^2$$

and the right side as

$$-\frac{c}{a} + \frac{b^2}{4a^2} = \frac{-4ac + b^2}{4a^2} = \frac{b^2 - 4ac}{4a^2}$$

we obtain the equation

$$\left(x + \frac{b}{2a}\right)^2 = \frac{b^2 - 4ac}{4a^2}.$$

Taking the square root of both sides, we have

$$x + \frac{b}{2a} = \pm\sqrt{\frac{b^2 - 4ac}{4a^2}} = \pm\frac{\sqrt{b^2 - 4ac}}{\sqrt{4a^2}} = \pm\frac{\sqrt{b^2 - 4ac}}{2a}$$

Subtracting $(b/2a)$ from both sides of this equation and using the term $2a$ as a common denominator, we obtain the solution for x as

$$x = \frac{-b \pm \sqrt{b^2 - 4ac}}{2a},$$

which is the quadratic formula.

Graphical Interpretation

The graph of the quadratic equation $y = ax^2 + bx + c$ is a parabola. When $a > 0$, the parabola opens upward. and the *minimum value* of y occurs at

$$x = -\frac{b}{2a}$$

If $a < 0$, the parabola opens downward and the *maximum value* of y occurs at

$$x = -\frac{b}{2a}$$

The value $x = -b/2a$ is referred to as the *x-component* of the parabola's vertex.

THE MATHEMATICS OF FINANCE

In this Chapter

One area of business where equations are used constantly is finance. Here many questions center on the relative value of different investments that return differing amounts of money at differing future times. For example, is a $5,000 investment now, that promises a $1,500 return for the next 5 years, better than a $4,500 investment now with a guaranteed return of $2,000 every other year for the next 8 years? Or, more personally, what investment plan should be undertaken if the goal is to accumulate $50,000 after 18 years to help pay for a college education. This chapter presents the mathematical underpinnings for these types of financial decisions.

4.1   SIMPLE INTEREST

Individuals organizations, businesses, and countries exchange their goods and services for the products of others. Bartering was one of the earliest means of establishing trade – a farmer and a weaver might exchange one bushel of corn for one wool scarf – but bartering soon gave way to currency, first in silver and gold coins and more recently script (paper money), as the primary unit of trade. Script itself has little intrinsic worth; the real value of money is its acceptance as a recognized unit of trade, just as bitcoin is being similarly recognized. With money as a medium, a bushel of corn worth $10 and a wool scarf worth $13 can be traded fairly, generally through a succession of wholesalers and distributors.

Money can either be saved, borrowed, and lent. Money is saved to buy consumer goods, such as television sets and iPhones; it is borrowed to finance purchases such as homes, cars, and college educations, and it is lent by banks and other financial institutions to make these purchases. Each dollar, pound, mark, shilling, yen, rubble, or peso that is lent or borrowed exact a charge or cost called ***interest***.

The amount of money lent or borrowed is called the ***principal***, usually denoted as ***P***, and the duration of the loan is its maturity, **denoted as *t***. In the simplest type of interest computation, the interest payment is directly proportional to the product of the principal and maturity. The constant of proportionality is the ***interest rate***, denoted as ***r***.

If we let ***I*** denoted the total interest, ***t*** the duration of the loan, and write ***r*** as a decimal value in terms of the same unit of time as t, then

$$I = P\,r\,t \qquad \text{(Eq. 4.1)}$$

Equation 4.1 is the *simple interest formula*.

Example 1 How much will a person have to pay in interest if they borrow $2,000 for 5 years at 3% simple annual interest?

Solution Using the formula $I = P\,r\,t$ (Eq. 4.1)

where

> $P = \$2,000$, $r = 0.04$ (that is, 4% written in decimal form), and $t = 3$
> $I = (\$2,000)(0.03)(5) = \$300$

With simple interest, interest payments are identical from one time period to the next because interest is always calculated on the *original* principal. Each year, the initial principal of $2,000 generates $60 of interest (3% of

$2,000). Over a five-year period, the total accumulated interest is $300 ($60 times 5), as shown in Figure 4.1.

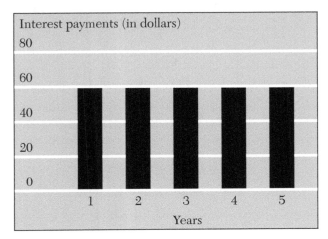

FIGURE 4.1 Three percent simple interest on $2,000.

Example 2 How much will a person have to pay in interest if they borrow $4,200 for 8 months at 9% simple annual interest?

Solution Using Equation 4.1 that $I = P r t$, with

$P = \$4,200$, $r = 0.09$ (that is, 9% written in decimal form), and $t = (8/12)$ years

$I = (\$4,200)\,(0.09)\,(8/12) = \$252.$

It is important to note that the value of r used in determining the total interest must have the same time dimension as the value of t. Thus, if r is given as a yearly rate, the duration, t, must be in years; similarly, if r is provided as a monthly rate, then t must be in months. Thus, in both Examples 1 and 2, t is used as a yearly value to match the rate, which is used as a yearly value. Thus, in Example 2, 8 months is equivalent to (8/12) years, which is the value of t used in the solution.

Example 3 How much will a person have to pay in interest if they borrow $5,000 for 1½ years at 1% simple monthly interest?

Solution Here, the interest rate r, is provided as a monthly value and the duration, t, as a yearly value. Thus, we must either convert the monthly rate to a yearly rate (accomplished by multiplying the monthly rate by 12), or convert the duration of 1½ years to 18 months (accomplished by multiplying the yearly duration by 12) Choosing to convert the duration to months and using the formula $I = P r t$ with

P = \$5,000, r = 0.01 (that is, 1% written in decimal form), and t = 18 months
$I = (\$5,000)(0.01)(18) = \$900$

The borrower in Examples 1, 2, and 3 must pay both the interest and principal at the conclusion of the loan. Similarly, an investor would expect to receive both her initial investment plus the interest at maturity. If we let A denote the value of a loan or investment at maturity, then

$$A = P + I = P + P r t = P(1 + rt) \qquad \text{(Eq. 4.2)}$$

The borrower in Example 1 will have to pay $A = P + I = \$2,000 + \$300 = \$2,300$ after five years. Similarly, the borrower in Example 2 must pay the lender $A = P + I = \$4,200 + \$252 = \$4,452$ at the end of eight months, and the borrower in Example 3 will have to pay back \$5,000 + \$900, or \$5,900 after 18 months.

Exercises 4.1

For Exercises 1 through 10, calculate (a) the amount of interest and (b) the total value of the investment (or loan) at maturity

1. \$1,000 at 6% simple annual interest for 4 years.

2. \$1,000 at 6% simple annual interest for 10 years.

3. \$1,000 at 1% simple monthly interest for 9 months.

4. \$2,500 at 7 ½% simple annual interest for 4 years.

5. \$2,500 at 7 ½% simple annual interest for 4½years.

6. \$2,500 at 0.5% simple monthly interest for 4¼ years.

7. \$3,200 at 10¼% simple annual interest for one-half-year.

8. \$25,000 at 4% simple annual interest for 3years.

9. \$6,825 at 7½% simple annual interest for 20years.

10. \$5,370 at 1.5% simple monthly interest for 1½ years.

11. An initial investment that paid 8% simple annual interest is worth $14,000 after five years. What was the initial principal invested?

12. An initial investment that paid 4½% simple annual interest is worth $1,4,02.50 after ten years. What was the initial principal invested?

4.2 COMPOUND INTEREST

Most interest payments, from common savings accounts in banks to unpaid balances on credit cards, involve compound interest calculations. Additionally, compound interest forms the primary foundation of finance, investment analyses, and modern portfolio theory. As such, compound interest is an essential topic required for delving into these more advanced financial topics.

The defining property of ***compound interest*** is that once interest is paid on the initial principal amount, the interest is immediately added to the principal. This new principal amount, which now consists of the original principal amount plus the interest, earns interest during the next time period. Thus, the interest earned in one time period (referred to as a compounding period) earns interest in succeeding periods; this is known as interest being paid on interest and is the defining characteristic of compound interest calculations.

As an example, consider the deposit of $2,000 in an account paying an annual interest rate of 3%, with compound interest computed and paid once a year. In the first year the principal earns 3% of $2,000 or $(0.03)($1,000) = 60. The new principal is now $2,060 (which is the original investment of $2,000 plus the $60 interest payment). Thus, the second year's interest payment is now based on this new amount, which becomes 3% of $2,060 or $61.80. This makes the balance at the end of the second year $2,121,80. Interest payments for the third year are now computed based on this new balance. The results of all interest computations through the fifth year have been collected in Table 4.1.

The yearly interest payments listed in Table 4.1 are illustrated in Figure 4.2. Notice that the interest payments for each year is greater than that of the previous year. The reason for this is that each year's interest is calculated on the *sum* of the initial principal *and* all prior interest payments (not just on the initial principal, as in simple interest calculations). Compare Figure 4.2 with the analogous simple interest payments shown in Figure 4.1.

TABLE 4.1 3% Interest Compounded Annually of $2,000.

1.	Original Investment	$2,000.00	$= P_0$
2.	Interest for the first year (3% of line 1)	$60.00	
3.	Principal during the second year (line 1 plus line 2)	$2,060.00	$= P_1$
4.	Interest for the second year (3% of line 3)	$61.80	
5.	Principal during the third year (line 3 plus line 4)	$2,121.80	$= P_2$
6.	Interest for the third year (3% of line 5)	$63.65	
7.	Principal during the fourth year (line 5 plus line 6)	$2,185.45	$= P_3$
8.	Interest for the fourth year (3% of line 7)	$65.55	
9.	Principal during the fifth year (line 7 plus line 8)	$2,251.00	$= P_4$
10.	Interest for the fifth year (3% of line 9)	$67.53	
11.	Principal at the end of the fifth year (line 9 plus line 10)	$2,318.53	$= P_5$

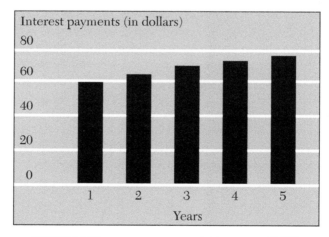

FIGURE 4.2 Three percent compounded interest on $2,000.

Obviously, we could continue Table 4.1 and find the principal at the end of any year. But this can be time-consuming, especially if we are interested in the principal after 25 or 30 years. Luckily, there exists a formula that allows us to calculate such principals with very little work, which we will present shortly.

To obtain the desired formula and understand its usage, reconsider Table 4.1. For notational simplicity, as listed in the last column of the table, the original principal amount is denoted as P_0. Continuing with this notation, P_1 denotes the principal after the first year, P_2 the principal after the second year, P_3 the principal after the third year, and so on. It follows from line 3 in Table 4.1 that the principal after the first year is:

$$P_1 = P_0 + (0.03)P_0 = P_0\,(1 + 0.03), \qquad \text{(Eq. 4.3)}$$

It follows from line 5

$$P_2 = P_1 + (0.03)P_1 = P_1\,(1 + 0.03) \qquad \text{(Eq. 4.4)}$$

Substituting for P_1 from Equation 4.3 into Equation 4.4, yield

$$P_2 = [P_0\,(1 + .03)](1 + 0.03) = P_0\,(1 + 0.03)^2 \qquad \text{(Eq. 4.5)}$$

Similarly,

$$P_3 = P_0\,(1 + 0.03)^3.$$

Clearly, these interest calculations can be carried on indefinitely. For example,

$$P_4 = P_0\,(1 + 0.03)^4,$$

and

$$P_5 = P_0\,(1 + 0.03)^5$$

This pattern can be generalized for any interest rate and time period to provide the desired final equation. Letting r denote the interest rate, n the number of interest payments that were made, and P_n the principal amount after the nth payment was made, we have,

$$\boldsymbol{P_n = P_0\,(1 + r)^n} \qquad \text{(Eq. 4.6)}[1]$$

Equation 4.6 enormously simplifies compound interest calculations. For annual interest payments, the principal amount at the end of the nth year (after which n interest payments have been made) is obtained by simply adding the interest rate to 1, raising this sum to the nth power, and multiplying this result by the original investment. To do this, a calculator that can raise a number to a power is needed to find the value of $(1 + r)^n$.

Example 1 Two thousand dollars is invested in an account that pays 3% interest compounded annually. Determine the balance after 5 years and the balance after 25 years.

---

[1] It is interesting to note that Equation 4.6 is an exponential equation having the form of Equation 3.9 in Chapter 3, with $a = P(0)$, $b = (1 + r)$, and the variable x replaced by n.

Solution Here $P = \$2{,}000$, $i = 0.03$, and we seek the principal after 5 years For these values, and setting $n = 5$ in Equation 4.4 we calculate

$$P_5 = (\$2{,}000)\ (\$2{,}000)\ (1 + 0.03)^5$$
$$= (\$2{,}000)\ (1.03)^5$$
$$= (\$2{,}000)\ (1.159265) = \$2{,}318.53.$$

Compare this value with the last dollar figure tabulated in Table 4.1.

To find the principal after 25 years, we set $n = 25$ in Equation 4.4. Again $P_0 = \$2{,}000$ and $i = 0.03$, hence,

$$P_{25} = (\$2{,}000)\ (1 + 0.03)^2$$
$$= (\$2{,}000)\ (1.03)^{25}$$
$$= (\$2{,}000)\ (2.093778) = \$4{,}187.56$$

It is certainly easier to obtain P_{25} this way than to continue Table 4.1 through another 20 years.

Compounding Periods

Interest rates are generally quoted on an annual basis but are typically compounded over shorter intervals of time. The annual rate, referred to as either the ***nominal interest rate*** or the ***stated interest rate*** is denoted by the symbol ***r***. The time between successive interval payments is called the ***compounding period***, or the ***period***, for short. The ***interest rate per period*** is denoted by the symbol ***i***; it is calculated by dividing the stated annual rate, ***r*** by the number of compounding periods in a year, which is denoted as N

$$\boldsymbol{i = r/N} \qquad\qquad \text{(Eq. 4.7)}$$

If the interest is compounded quarterly, then N is 4 (there are four-quarters in a year, and $i = r/4$. For interest compounded semiannually, $N = 2$ and $i = r/2$; for interest compounded monthly $N = 12$ and $i = r/2$; and for interest compounded. Weekly $N = 52$ and $i = r/52$. If no compounding period is stated, compounding periods are assumed to be annual and $i = r$. This information is summarized in Table 4.2, which lists the most commonly used compounding periods and the interest rates that apply to them, where i is the stated annual interest rate.[2]

---

[2] Continuous compounding is presented in Section 4.9.

TABLE 4.2 Calculating the Interest rate per Compounding Period.

Compounding Period	Compounding Periods per Year	Interest Rate per Compounding Period (r = the stated annual rate)
Annually	1	$i = r$
Semiannually	2	$i = r/2$
Quarterly	4	$i = r/4$
Monthly	12	$i = r/4$
Daily	365	$i = r/365$

As seen in the last column of Table 4.2, the *interest rate per compounding period* is the annual rate divided by the number of compounding periods in a year.

Equation 4.7 remains valid for all the compound periods listed in Table 4.2, as long as we realize that i signifies the interest rate per compound period, and P_n is the balance after n compound periods. For example, if the interest is 2% compounded quarterly, $i = 0.02/4 = 0.005$, which is the interest rate per quarter. Also, P_{10}, for example, denotes the principal after 10 compounding periods which, in this case, is 10 quarters and corresponds to 2½ years.

Example 2 Ten thousand dollars is invested in an interest-bearing account that pays 4% interest compounded quarterly. Determine the balance after 5 years.

Solution Because interest is paid quarterly, we take one-quarter of a year as our basic time period. Then, the balance after 5 years is given by the balance after 20 quarters, which is P_{20}. The rate applied each quarter is the annual rate divided by 4 or $0.04/4 = 0.01$. Using Equation 4.4 with $n = 20$, $i = 0.01$, and $P_0 = \$10,000$, we obtain

$$P_{20} = (\$10,000)(1 + .01)^{20}$$
$$= (\$10,000)(1.01)^{20}$$
$$= (\$10,000)(1.220190) = \$12,201.90.$$

Example 3 Ten thousand dollars is invested in a savings account that pays 4% interest compounded semiannually. Determine the balance after 5 years.

Solution Because interest is paid semiannually, we take one-half of a year as our basic time period. Accordingly, the balance after 5 years is given by the balance after 10 half-years, or P_{10}. The applicable interest rate per compounding period is the annual rate divided by 2 or $0.04/2 = 0.02$. Using Equation 4.4 with $n = 10$, $i = 0.02$, and $P_0 = \$10,000$, we obtain[3]

$$P_{10} = (\$10,000)\,(1 + .02)^{10}$$
$$= (\$10,000)\,(1.02)^{10}$$
$$= (\$10,000)\,(1.2189944) = \$12,189.94.$$

Example 4 Ten thousand dollars is invested in an investment account that pays 2% interest compounded monthly; determine the balance after 5 years.

Solution Here $i = 0.02/12$ and we seek P_{60}, the balance after 60 months or 5 years. Using Equation 4.4, we obtain

$$P_{60} = (\$10,000)\,(1 + 0.02/12)^{60}$$
$$= ((\$10,000)1 + 0.001667)^{60}$$
$$= ((\$10,000)1\,.001667)^{60}$$
$$= (\$10,000)\,(1.10507895 = \$11,050.79.$$

Exercises 4.2

1. For $2,000 deposited in an account for four years that yields 2% annual interest compounded annually:

   a. Determine the values of i, n, and P_0 that would be used to determine the amount in the account at the end of the fourth year.

   b. Determine the balance in the account at the end of the fourth year.

2. For $2,000 being deposited in account for 5 years that yields 3% annual interest compounded monthly:

   a. Determine the values of i, n, and P_0 that would be used to determine the amount in the account at the end of the fifth year.

   b. Determine the balance in the account at the end of the fifth year.

3. Ms. Brown borrows $2,500 from a friend who charges 4% interest compounded annually. Determine her debt after 3 years.

4. Redo Exercise 3 with the interest compounded quarterly.

5. Ms. Brown invests $2,500 in a venture that pays 5% interest compounded quarterly. Determine her balance after 3 years.

---

[3] Generally, six or more decimal places are retained in interest rate calculations for each $1,000 of principal . More complicated financial instruments typically require retaining at least 10 significant digits of accuracy in all intermediate calculations.

6. Redo Exercise 5 with the interest compounded semiannually.

7. Mr. Johnson deposits $1,000 in an account that pays 2.5% interest compounded annually. How much will he have after 25 years?

8. Redo Exercise 7 with the interest compounded semiannually.

9. Determine the balance after 3 years resulting from $2,900 being deposited in a savings account that pays 1.5% interest compounded monthly.

10. Determine the balance after 1 year resulting from $3,500 being deposited in a savings account that yields 2.5% interest compounded daily.

11. Redo Exercise 10 for the balance after 4 years.

12. Some institutions use an *approximate year* rather than a calendar year for certain interest computations. In this method, every month is assumed to have exactly 30 days, resulting in an approximate year of 360 days. Using an approximate year, set up and determine the balance after 3 years for an initial deposit of $2,500 if the interest rate is 3% compounded daily.

4.3 LUMP-SUM FUTURE AND PRESENT VALUES

A *lump-sum* is a dollar amount made as a one-time single payment. Examples of lump-sum payments are an initial deposit, a single one-time dollar investment, or a final, single loan repayment. Equation 4.6 relates a lump-sum principal amount at two points in time–the present, when the principal is first deposited, and its value in the future. The reason these values differ is due to the interest that is earned.

In this section, we rewrite and use Equation 4.6 in two different ways to emphasize this time relationship. To do this, we will use standard financial notation that emphasizes the two unique usages. The first usage emphasizes determining P_n, the future value of an initial principal amount, given that we know P_0. The second usage emphasizes the equation's use in determining the initial amount deposited, that is P_0, given that we know P_n, its future value. In financial applications, this second usage is typically much more important and the key to comparing investment alternatives.

Lump-Sum Future Values

For convenience, we first reproduce Equation 4.6 as Equation 4.8, so that we can rewrite it using standard financial notation. The advantage of this new notation is that it clearly relates the values of the principal amounts at two differing points in time, the present and the future.

$$P_n = P_0(1 + i)^n \qquad \text{(Eq. 4.8)}$$

Financially, P_0, the initial principal, is referred to as the *present value of the principal*, or *present value*, for short. The notation used for this quantity is **PV**. Similarly, P_n, which denotes the value of this principal sometime in the future, is referred to as the *future value of the principal*, or *future value*, for short. The notation used for this quantity is **FV**. Note that this notation emphasizes what these quantities actually represent in time (now and in the future), as opposed to their strictly mathematical relationship.

Using this new notation, Equation 4.8 is rewritten as:

$$FV = PV\,(1 + i)^n \qquad \text{(Eq. 4.9)}$$

Figure 4.3 illustrates the relationship provided by Equation 4.9.

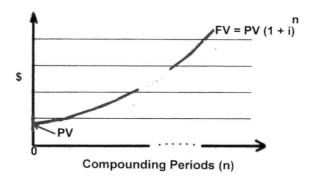

FIGURE 4.3 The Time value of money.

Mathematically, Equations 4.8 and 4.9 are identical. The notation used in Equation 4.9, however, reinforces the time dependences and is the predominant one used in almost all financial applications. Thus, given the present value of a single investment amount, denoted as *PV*, and the interest rate per compound period, *i*, Equation 4.9 is used to calculate the future value of this investment *n* compound periods later, exactly in the same manner as

Equation 4.6 was used in the prior section. The relationship illustrated in Figure 4.3 is frequently referred to as ***the time value of money***; that is, the value of any money received is relative to the time period in which it is received. Another way of looking at this is that a dollar received a year from now is not equivalent to a dollar received today. The reason is that a dollar received today can be placed in an interest-earning account, so that it will be worth more in the future.

Equation 4.9 is particularly important when there are several different ways to invest the same amount of money, and we must determine which one of the choices will be the most profitable. For example, suppose we have $10,000 in available cash and are invited to invest this money in a land venture with an expected return of $12,000 in 5½ years. If we decide against the land venture, we can deposit our money in an account with a guaranteed interest of 3% per year compounded quarterly. What should we do?

To answer this question based on monetary values, we must compare the amounts each investment will return at the *same point in time,* in this case after 5 ½ years. The future value of the land venture is fixed at $12,000. What is the future value of the money if it is deposited in the interest-bearing account?

Because the interest is compounded quarterly, the applicable interest rate, i is $0.03/4 = 0.0075$ per quarter. Thus, the present value of our investment, PV, is $10,000, and we seek the future value, FV, after $n = 22$ quarters. Using Equation 4.9, the future value of the investment is calculated as:

$$FV = (\$10,000)\,(1 + 0.0075)^{22} = (\$10,000)\,(1.1787667) = \$11,787.67.$$

Clearly the land venture, on a purely monetary basis, is a more profitable investment by $212.33.[4]

Example 1 Mr. James' barber would like Mr. James to lend him $6,000 for the modernization of his barber shop. The barber promises to pay Mr. James $6,500 at the end of 2 years. How does this investment compare with investing the same money in corporate bonds for 2 years at 4% compounded semiannually?

Solution At the end of 2 years, the future value of the barber shop investment is $6,500. To determine the future value of $6,000 in corporate bonds, we use Equation 4.9 with $i = 0.04/2 = 0.02$, $n = 4$, and $PV = \$6,000$. Then,

$$FV = (\$6,000)\,(1 + 0.02)^{4} = (\$6,000)\,(1.082432) = \$6,494.59.$$

---

[4] This does not consider the relative risk of each investment, which is referred to as the *credit risk.*

Therefore, from a strictly monetary standpoint, the barber shop invest-ment offers the greater future value at the end of 2 years by $5.41. From an investment standpoint these two returns are essentially the same, so from a strictly financial point of view, the decision, should be made on risk – that is, the investment with the least risk in terms of Mr. James actually receiving the final payment is preferable.

Example 2 Ms. Everet has $10,000 to deposit. One bank offers 1.5% inter-est compounded annually, while a second bank offers 1.25% interest com-pounded monthly. Which bank should she choose if she wants the greatest return after 4 years?

Solution To determine the future value of $10,000 at the first bank we use Equation 4.9 with $i = 0.015$, $n = 4$, and $PV = \$10,000$. Then

$$FV = (\$10,000) (1 + 0.015)^4 = (\$10,000 (1.061364) = \$10,613.64.$$

To determine the future value of $1,000 at the second bank, we use Equa-tion 4.9 with $i = 0.0125/12$, $n = 48$, and $PV = \$10,000$. Then,

$$FV = (\$10,000 (1 + 0.0125/12)^{48} = (\$10,000 (1.051244) = \$10,512.44.$$

Ms. Everet will do better at the first bank.

Lump-Sum Present Values

Equation 4.9 gives the future value of lump-sum amount in terms of the pre-sent value. In a variety of financial situations, the amount of money either needed in the future, or known to be available in the future is given, and its equivalent value today is required. For example, one might desire a specific amount of money in the future for college tuition or for a down payment on a house and want to know how much money this represents today. Or the return in the future from an investment might be known, and its equivalent value today is desired.

Solving for present values when future values are known is actually very common in financial decision making; much more common, in fact, then solv-ing for future values given present values. For these situations, Equation 4.9 is rewritten to more easily solve for the present values. Dividing both sides of Equation 4.10 by $(1 + i)^n$, we obtain

$$PV = FV / (1 + i)^n$$

Which can be rewritten as:

$$PV = FV \left(1 + i\right)^{-n} \qquad\qquad \text{(Eq. 4.10)}$$

Equation 4.10 gives the present value of a lump-sum amount of money in terms of its future value.

Example 3 Determine the present value of $15,000 due in 5 years at 5% compounded annually.

Solution Here we have no money available in the present but we will receive $15,000 in the future. Therefore $FV = \$15,000$. Substituting this value into Equation 4.10, with $n = 5$, and $i = 0.05$ yields

$$PV = (\$15,000) \left(1 + 0.05\right)^{-5} = (\$15,000) \, (0.783526) = \$11,752.89.$$

The present value is the amount required *now* that, dollar for dollar, is equivalent to the stated future value. It follows from Example 3 that $11,752.89 now is equivalent to $15,000 in 5 years if the funds are placed in an account that pays 5% interest compounded annually. In other words, if $11,752,89 is invested today at 5% compounded annually, it will grow to $15,000 in 5 years.

Example 4 Mr. Kakowski's bank is offering 10-year secured certificates of deposit at 2% interest compounded quarterly. Certificates can be purchased in $100 amounts with a minimum deposit of $1,000. How much should Mr. Kakowski invest now if he wants a final return of $20,000 in ten years?

Solution We are given $FV = \$20,000$, $n = 40$ and $i = 0.02/4 = 0.005$, and we seek the PV. Substituting these values into Equation 4.10, we obtain:

$$PV = (\$20,000) \left(1 + 0.005\right)^{-40} = (\$20,000) \, (0.81913886) = \$16,382.77$$

Because certificates can be purchased in $100 lots only, Mr. Kakowski would need to invest $16,400 now.

Comparing Investment Alternatives

Often one has to choose between several different investment opportunities, each having a known future value. For example, is an investment that returns $50,000 at the end of 8 years better than one that returns $43,500 at the end of 5 years? In problems such as these, the potential profits of each investment must be compared at the same point in time before an appropriate decision can be made. Because the present is a common point in time for all investments, no matter when in the future they provide a return, present values are almost always used in making such comparisons. In doing so, Equation 4.10 is used to convert all future values, no matter when they occur, to their present values.

Example 5: Mr. Kingsley plans to sell his bakery and retire. Two of his employees wish to buy the bakery, but they do not have any immediate cash. They expect to make money from operating the bakery, so each makes an offer. Employee A wants the business for $50,000 payable at the end of 8 years. Employee B wants the business for $44,500 due at the end of 5 years. Which offer is better at an interest rate of 4% per year?

Solution Because each offer provides a return at a different date in the future, simply comparing the $50,000 to the $44,500 is not valid. To make a valid comparison, both offers must be evaluated at the same point of time. By convention, this time is taken to be the present. Thus, we must translate each offer to its present value and then compare them.

Employee A's offer:

$$PV = (\$50,000)\,(1 + 0.04)^{-8}$$
$$= (\$50,000)\,(0.730690) = \$36,534.50$$

Employee B's offer:

$$PV = (\$44,500)\,(1 + 0.04)^{-5}$$
$$= (\$44,500)\,(0.821927) = \$36,575.75$$

Because the present value of employee B's offer is higher, it is the better offer.[5] Here the present value represents the equivalent cash settlement *now*. That is, $50,000 due in 8 years is equivalent to receiving $36,534.51 today, assuming that the funds can be invested at 4%. Similarly, $44,500 due in 5 years is equivalent to $36,575.75 now; again assuming a 4% investment rate. Thus, what might appear to be the lower offer ($44,500) is actually, the better offer.

Example 6: The Die-Cast Corporation has three offers for its die-casting equipment. The first buyer is willing to purchase the equipment for $50,000, payable at the end of 8 years. The second buyer is willing to pay $39,000, consisting of an immediate payment of $14,000 now and $25,000 due in 6 years. The third buyer will purchase the equipment for $35,000 payable immediately. Determine the best offer for the equipment assuming all potential purchasers can meet their obligations and the Die-Cast Corporation can deposit all money received in an interest-bearing account that pays 5% interest annually.

---

[5] Again, this is strictly on a monetary basis, and does not taking into account any other factors, such as the relative risk between the two employees.

Solution Because each offer matures at a different date, we first compute their respective monetary values at the present time (PVs).

First buyer:

$$PV = (\$50{,}000) \, (1 + 0.05)^{-8}$$
$$= (\$50{,}000) \, (0.676839) = \$33{,}841.96$$

Second buyer:

$$PV = (25{,}000) \, (1 + 0.05)^{-6} + \$14{,}000$$
$$= (25{,}000) \, (0.746215) + \$14{,}000$$
$$= \$18{,}655.38 + \$14{,}000 = \$32{,}655.38$$

Third buyer:

$$PV = \$35{,}000.$$

The third offer is best, because its present value is the highest.

Exercises 4.3

Future Value Problems:

1. Determine the future value of $5,000 after 10 years if it is deposited in an account that pays 4% compounded.

   a. annually

   b. semiannually

   c. quarterly

   d. daily (assume each year consists of 365 days)

2. Determine the future value of $12,000 after 5 years at 2% interest compounded.

   a. annually

   b. semiannually

3. Determine the future value of $12,000 after 7 years at 5% interest compounded quarterly.

4. Determine the future value of $8,000 after 10 years at 1.5% interest compounded semiannually.

5. A man has $1,000 to deposit. Should he put it in a bank offering 2% interest compounded quarterly or one offering 4% interest compounded annually?

6. Ms. Field's financial advisor has recommended that she invest $20,000 in a new housing development with an anticipated return of $28,000 in 6 years. The advisor claims that this is a better investment than investing her money in an account that pays 6% interest compounded annually. Is the advisor correct?

7. Ms. Wilson has $2,000 to invest. Either she can deposit this money in a time savings plan that will pay 1.5% annual interest or she can lend the money to a friend who will repay her $750 at the end of each year for the next 3 years. Which opportunity is more profitable assuming that interest rates remain at their current level?

8. For Exercise 7, determine the more profitable opportunity if the interest rate is 8% annually.

Present Value Problems:

9. Determine the present value of $15,000 due in 8 years at

   a. 2% interest compounded annually.

   b. 4% interest compounded annually.

   c. 8% interest compounded annually.

   d. Based on the present values determined in parts a. through c. what can you say is the relationship between present values and the interest rate?

10. With the birth of their son, the Boswells decide to deposit a sum of money in an account paying 3% annual interest compounded annually. Their objective is to accumulate enough money to provide the son with $20,000 at his 18th birthday. Determine the amount that they should invest now in order to meet their objective.

11. How much money should be deposited in a Certificate of Deposit that pays 4% compounded semiannually if the desired objective is $10,000 after 4½ years?

12. With the birth of their daughter, the Tucks decides to place a sum of money in an account which yields 3.5% compounded semiannually. If the objective is to accumulate $30,000 for their daughter's twenty-first birthday, determine the amount of the deposit. How much money will be available if the Tuck's give the money to their daughter on her twenty-fifth birthday?

Comparing Investment Alternatives

13. Dr. Baxter has $10,000 for investment purposes. She can put it into a friend's business with an expected return of $12,000 in 3 years, or she can invest it in an account that pays 4% interest compounded quarterly. Which opportunity is the most profitable?

14. Mr. Jones has two buyers for his business. Buyer A will pay $10,000 immediately and another $25,000 in 7 years. Buyer B will pay $8,000 immediately and another $27,000 in 5 years. Which is the better offer if the interest rates are 3% compounded annually?

15. A small business owner has three buyers for his business. Buyer A will pay $20,000 now and another $5,000 at the end of 4 years. Buyer B will pay $15,000 now and another $10,000 at the end of 3 years. Buyer C will pay $10,000 now and another $18,000 at the end of 6 years. Which is the best offer if the interest rate is 4% compounded annually?

16. An individual has three possible opportunities for investing the same amount of money. The first will return $8,000 in 4 years, the second will return $7,000 in 2 years, and the third will return $10,000 in 7 years. Which opportunity is the most attractive if the interest rate is 2% compounded annually?

17. Redo Exercise 16 with an interest rate of 6% compounded annually.

18. Mr. Smith has two buyers for his small business. Buyer A will pay $15,000 immediately and another $30,000 in 3 years. Buyer B will pay $9,000 immediately and another $35,000 in 2 years. Which is the better offer if the interest rates are 3% compounded annually?

Interest Rate Problems

19. Solve Equation 4.9 for i and show that

$$i = \left(\frac{FV}{PV}\right)^{1/n} - 1.$$

20. Using the results from Exercise 18, determine the annual interest rate required to convert $1,000 to $1,350 in 3 years.

21. Using the results from Exercise 18, determine the annual interest rate required to convert $10,000 to $15,000 in 10 years.

22. Using the results from Exercise 18, determine the annual interest rate required to double an investment after 10 years.

4.4 CASH FLOW NET PRESENT VALUES (NPVs)

In Sections 4.1 through 4.3, we concerned ourselves with single lump-sum payments. Thus, we either calculated the future value of a lump-sum invested now, or we calculated the present value of a lump sum payment to be made in the future. In this section we consider investments consisting of a *set* of payments due at *different* times, a situation known as a **cash flow**.

As an example of a cash flow, consider an investment that returns $500 in 1 year, another $300 in 3 years, and a final $400 in 4 years, with interest rates of 5% compounded annually. What is the present value of such an opportunity? That is, what is the cash equivalent now of the entire transaction?

A simple approach is to compute the present value of each of the individual payments using Equation 4.10, repeated below as Equation 4.11 for convenience, and then sum the individual present values to obtain the present value of the entire cash flow.

$$PV = FV(1 + i)^{-n} \qquad \text{(Eq. 4.11)}$$

Example 1 Compute the present value of the cash flow that returns $500 in 1 year, another $300 in 3 years, and a final $400 in 4 years, with interest rates of 5% compounded annually.

Solution The first payment of $500 is due in 1 year. The present value of this amount, computed using Equation 4.11 is

$$PV_1 = (\$500)(1 + 0.05)^{-1} = \$476.19.$$

The second payment of $300 is due in 3 years. Again using Equation 4.11, we find its present value as:

$$PV_2 = (\$300)(1 + 0.05)^{-3} = \$259.15.$$

Similarly, the present value of the last payment is

$$PV_3 = (\$400)(1 + 0.05) = \$329.08.$$

Summing these three present values, we obtain the present value of the entire investment as:

$$PV = PV_1 + PV_2 + PV_3 = \$476.19 + \$259.15 + \$329.08 = \$1,064.42.$$

In most present-value problems, a *time diagram* illustrating the contributions to the total present value from the individual payments is helpful. The time diagram for the cash flow given in Example 1 is shown as Figure 4.4.

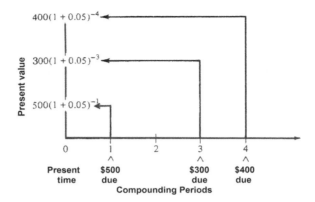

FIGURE 4.4 An example of cash flows.

This approach of summing individual present values is the general procedure for calculating present values for all cash flows; simply use Equation 4.11 to find the present value of each payment and then sum the results. One modification, however, is usual. Rather than finding the present value, it is more common to find the **Net Present Value (NPV),** which is the present value *minus* the cost of the investment. The net present value measures the additional money over and above the cost of the investment, which would have to be deposited in a bank to equal the returns provided by the investment. The net present value, which is also referred to as the *discounted cash flow* (DCF – this is an older term) is obtained as:

$$NPV = \sum PV \text{ of all cash inflows} - \sum PV \text{ of all cash outflows} \quad \text{(Eq. 4.12)}$$

Example 2 Ms. Tilson invests $1,000 in a friend's business. In return, her friend promises to pay Ms. Tilson $500 in a year, another $300 in 3 years, and an additional $400 in 4 years. Determine the net present value of this investment if the current interest rate is 5% compounded annually.

Solution We previously calculated the present value of this investment as $1,064.42. Because the investment costs $1,000, the net present value is

$$NPV = \$1,064.42 - \$1,000.00 = \$64.42.$$

Example 3 Determine the net present values of the following two investments, using an interest rate of 2% per annum compounded quarterly. The first is a $5,000 investment which returns $1,500 every year for the next 5 years while the second is a $4,500 investment which returns $2,000 every other year for the next 8 years.

Solution First, because the interest is compounded quarterly, we take one-quarter of a year as our basic time unit and $i = 0.02/4 = 0.005$. The time-diagrams for both investments are given in Figures 4.4 and 4.5, respectively.

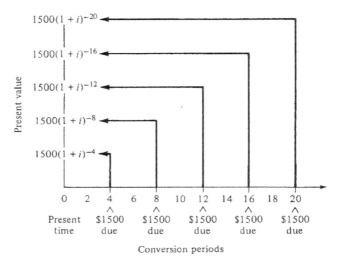

FIGURE 4.5 An example of cash flows.

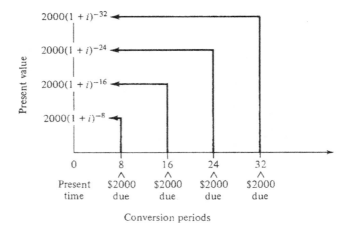

FIGURE 4.6 An example of cash flows.

The individual present values for the first investment are:

$$(\$1,500)\,(1+0.005)^{-4} = \$1,470.37$$
$$(\$1,500)\,(1+0.005)^{-8} = 1,441.33$$
$$(\$1,500)\,(1+0.005)^{-12} = 1,412.86$$
$$(\$1,500)\,(1+0.005)^{-16} = 1,384.95$$
$$(\$1,500)\,(1+0.005)^{-20} = 1,357.59$$
$$\text{Total} = \$7,067.10$$

Because the cost of participating in this investment is $5,000, the net present value is:

$$NPV = \$7,067.10 - \$5,000 = 2,067.10.$$

The individual present values for the second investment are:

$$(\$2,000)\,(1+0.005)^{-8} = \$1,921.77$$
$$(\$2,000)\,(1+0.005)^{-16} = 1,846.60$$
$$(\$2,000)\,(1+0.005)^{-24} = 1,774.37$$
$$(\$2,000)\,(1+0.005)^{-32} = 1,704.97$$
$$\text{Total} = \$7,247.71$$

Because the cost of participating in the second investment is $4,500, the net present value is:

$$NPV = \$7,247.71 - \$4,500 = \$2,747.71.$$

As the second investment has a higher net present value, it is the more attractive investment. Again, this is based strictly on monetary analysis and assumes the credit risk of both investments are equal.

The type of cash flow illustrated in Figures 4.5 and 4.6, where the same monetary amount is received at equal periods of time, occurs quite frequently in a number of practical investment situations. For a specific type of cash flow, where the compound periods and payment dates coincide, single formulas exist for directly computing both the cash flows' present and future values. We turn to these cash flows in the next section.

Exercises 4.4

1. Mr. Samuels is invited to invest $2,500 in a venture that will return $750 in 1 year, another $1,100 in 2 years, and a final $2000 in 4 years. Determine whether or not this is a profitable investment if current interest rates are 2% per annum compounded annually.

2. The guaranteed returns of three different investment opportunities are listed in Table 4.3. Which one is the most desirable at an annual interest rate of 3% compounded quarterly if each investment requires an initial cash outlay of $2,400?

TABLE 4.3

	In 1 year	In 2 years	In 3 years	In 4 years	In 5 years
	Guaranteed Returns				
Investment A	$1,000	$1,000	$1,000	$1,000	$1,000
Investment B	0	$2,500	0	0	$2,500
Investment C	$600	$800	$2,200	$800	$600

3. Determine the net present value of an opportunity that costs $2,900 and will return $500 in half a year, $1,000 in a year and a quarter, and $2,000 in 3 years, if the current interest rate is 2.5% per annum compounded monthly.

4. A man can invest $50,000 now and receive $12,000 at the end of each six months for the next 2 years plus an additional $12,000 at the end of the second year, or he can invest $70,000 now and receive $21,000 at the end of each six months for the next 2 years. Which opportunity is the most attractive at 4% interest per annum, compounded semiannually?

5. Carolyn's friend invited her to invest $5,000 in a venture that will return $800 in one year, another $1,500 in two years, and a final investment $3,500 in three years. Determine whether or not this is a profitable investment if the current rates are 2.75% per annum compounded annually.

6. Determine the present value of an opportunity that will return $600 in half a year, $1,200 in a year and three quarter, and $1,900 in three years if the current interest rate is 3% per annum compounded quarterly.

7. For Exercise 6, determine the more profitable opportunity if the interest rate is 8% Compounded annually.

8. Mr. Bates has $5,000 to invest. Determine the NPV of his investment if he lends the money to his neighbor, who repays him $800 at the end of each year for the next 4 years. Use an annual interest rate of 4.15%.

4.5   ORDINARY ANNUITIES

The present and future values of a cash flow can always be determined by calculating the present or future values, respectively, of each individual payment using the appropriate equation – either Equation 4.9 or 4.10, repeated below as Equations 4.13 and 4.14 for convenience – and then summing the results.

$$FV = PV (1 + i)^n \qquad \text{(Eq. 4.13)}$$

or

$$PV = FV (1 + i)^{-n} \qquad \text{(Eq. 4.14)}$$

For a specific type of investment, however, known as an *annuity*, the final sum can be calculated using a single formula.

Definition 4.1 An *annuity* is a set of equal payments made at equal intervals of time.

Car loans, mortgages, life insurance premiums, social security payments, and bond coupon payments are all examples of annuities. In each, one party, be it an individual, company, or government, pays to another party a set of equal payments, called *periodic installments* or *payments*, denoted as *PMT*, at equal periods of time, called the *rent period, payment period, payment interval, or compounding period*. Each of these terms can be used interchangeably.

Annuities are classified as either *ordinary* or *due*. With an *ordinary annuity*, payments are made at the end of each payment period, whereas with an *annuity due*, payments are made at the beginning of each period. Examples of ordinary annuities are car loan payments, mortgages, and bond coupon payments. Examples of annuities due are typically savings plans, pension plans, and lottery winnings that are paid over time.

An annuity is *simple* if the compounding period at which interest is paid coincides with the payment dates. In this section, we consider simple ordinary annuities; simple annuities due are presented in Section 4.7.

Present Value of an Ordinary Annuity

Figure 4.7 illustrates a time diagram for a typical ordinary annuity, with *PMT* dollars due each compound period for the next *n* periods, and where the present value of the complete cash flow is desired.

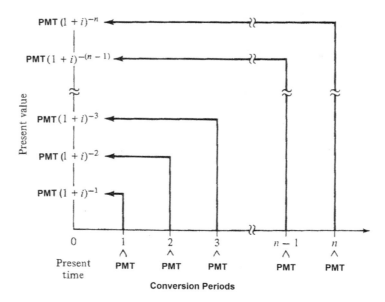

FIGURE 4.7 Present value of an ordinary annuity.

As shown in Figure 4.7, the present value of the ordinary annuity illustrated is given by:

$$PV = PMT(1+i)^{-1} + PMT(1+i)^{-2} + PMT(1+i)^{-3} + \ldots$$
$$+ PMT(1+i)^{-(n-1)} + PMT(1+i)^{-n}$$

Note that each term represents the present value of *one* of the future payments. In particular, $PMT(1+i)^{-1}$ is the present value of the first payment, $PMT(1+i)^{-2}$ is the present value of the second payment, and $PMT(1+i)^{-n}$ is the present value of the last payment. Factoring PMT from these terms, we have

$$PV = PMT[(1+i)^{-1} + (1+i)^{-2} + (1+i)^{-3} + \ldots + (1+i)^{-(n-1)} + (1+i)^{-n}]$$

The terms in brackets are a geometric series, whose sum can be written as:

$$\left[\frac{1-(1+i)^{-n}}{i} \right]$$

Thus, the Present Value becomes

$$PV = PMT\left[\frac{1-(1+i)^{-n}}{i} \right] \qquad \text{(Eq. 4.15)}$$

If the net present value is required, which takes into account the amount paid to receive the stream of income shown in Figure 4.7, all that is required is to subtract out this initial payment, denoted as C_0. That is,

$$NPV = PMT\left[\frac{1-(1+i)^{-n}}{i}\right] - C_0 \qquad \text{(Eq. 4.16)}$$

Example 1 Determine the value of

$$\left[\frac{1-(1+0.07)^{-15}}{0.07}\right]$$

Solution Using a calculator, the value of this term is determined as

$$\left[\frac{1-(1+0.07)^{-15}}{0.07}\right] = \left[\frac{1-0.36244602}{.07}\right] = \frac{.63755398}{.07} = 9.107914$$

Example 2 Determine the value of

$$\left[\frac{1-(1+0.005)^{-120}}{0.005}\right]$$

Solution

$$\left[\frac{1-(1+0.005)^{-120}}{0.005}\right] = \left[\frac{1-.54963273}{0.005}\right] = \left[\frac{.45036727}{0.005}\right] = 90.073453$$

Example 3 Determine the present value of an investment that returns \$25 every month for the next 4 years at 2% interest compounded monthly.

Solution Using Equation 4.15 with $i = 0.02/12$, $n = 48$, and $PMT = \$25$ we have

$$PV = \$25\left[\frac{1-\left(1+\dfrac{0.02}{12}\right)^{-48}}{\dfrac{0.02}{12}}\right] = \$25\left[\frac{1-0.92317782}{\dfrac{0.02}{12}}\right]$$

$$= \$25\left[\frac{0.07682218}{0.02/12}\right] = \$25[46.093308] = \$1,152.33$$

Example 4 Determine the net present value of an investment costing $900 that will return $45 every quarter for the next 5 and ½ years if current interest rates are 4% per annum compounded quarterly.

Solution Because interest rates are compounded quarterly, we take one-quarter of a year as our basic time period. Then $i = 0.04/4 = 0.01$ Here $PMT = \$45$, $n = 22$, and $C_0 = \$900$. Using Equation 4.16, we calculate

$$NPV = \$45 \left[\frac{1 - \left(1 + \frac{0.04}{4}\right)^{-22}}{\frac{0.04}{4}} \right] - \$900$$

$$= \$45 \left[\frac{1 - 0.80339621}{\frac{0.04}{4}} \right] - \$900$$

$$= \$45 \left[\frac{0.19660379}{0.01} \right] - \$900$$

$$= \$45 [19.660379] - \$900$$

$$= \$884.72 - \$900 = -\$15.28,$$

A negative NPV indicates the investment results in a loss. Here, putting the $900 in a bank at 4% compounded quarterly for 5½ years yields $900 $(1 + .04/4)^{22} = \$900 \,(1.244716) = \$1,120.24$. Obviously, it is more profitable *not to* partake in the investment and simply deposit the available cash in an account paying the stated rate of 4%.

Equations 4.15 and 4.16 are relatively easy to use when they apply; however, one must be careful to apply them correctly. First, *these equations are valid only if the payment dates and the compounding dates coincide.* This is not the case, for example, in Example 2 of the prior section, where the first investment has four compounding dates between successive payment dates, and the second investment has eight compounding dates between successive payment dates. Second, *the payments must all be equal.* Third, *one must remember that i denotes the rate per compounding period,* which generally is not the annual rate, and that *n* denotes the number of compounding periods in the investment, which usually differs from the number of years of the investment. Nonetheless, when Equations 4.15 and 4.16 are applicable, they save a good deal of work. It even can be combined with Equation 4.9 to determine net present values of investments involving both time and single lump sum payments.

Example 5: Dr. Ericson plans to invest in a corporate bond that returns $20 in dividends every half-year for the next 20 years plus an additional $1,000 at the end of the twentieth year. Determine the present value of this investment if interest rates are 3% per annum compounded semiannually?

Solution We separate the problem into two parts: one involving the $20 payments, and the other involving the $1,000 final payment. We find the total present value by calculating the present value of each part of the investment separately, and then summing to obtain the total present value.

The dividend payments represent an ordinary annuity.[6] Thus, the present value of the dividend cash flow is obtained from Equation 4.15. Here $PMT = 20$, $n = 40$, and $i = 0.03/2 = 0.015$. Therefore,

$$PV = \$20 \left[\frac{1 - \left(1 + \dfrac{0.03}{2}\right)^{-40}}{\dfrac{0.03}{2}} \right] = \$20 \left[\frac{1 - 0.55126232}{0.015} \right]$$

$$= \$20 \left[\frac{0.44873768}{0.015} \right] = \$20 [29.91584533] = \$598.32$$

The present value of the $1,000 lump sum payment due in 20 years or 40 half-years can be obtained directly from Equation 4.10.

$$PV = (1 + 0.015)^{-40} (\$1,000) = (0.55126232)(\$1,000) = \$551.26.$$

Therefore, the total present value of the entire transaction is

$$PV = \$598.32 + \$551.26 = \$1,149.58$$

Example 6: Redo Example 5 if the interest rate is 6%

Solution Applying the same calculations as in Example 5, the total present value of the entire cash flow ($20 annuity payments plus the final $1,000) payment is:

$$PV = \$462.30 + \$306.56 = \$768.86$$

Note that at the higher interest rate, the present value is less than it is at the lower interest rate.

---

[6] This is true for the majority of all corporate and government issued bonds.

Commentary: Financially, the present value of a bond represents its price, and what is being paid for is a stream of income from the coupons and the return of the initial principal. Examples 5 and 6 illustrate the fundamental inverse relationship between bond prices and interest rates; that is, *if interest rates rise the price of a bond falls, and if interest rates fall the price of the bond rises.* The reason is that the coupon payment is fixed when the bond is issued. Because the coupon payment amounts are fixed, the interest rate an investor receives when the bond is subsequently traded, is adjusted by paying more or less for the fixed income stream, depending on the current interest rate.

A second adjustment made to a bond's price occurs when the bond is purchased on a non-coupon payment date. In these cases, which form the majority of bond trading, the price of the bond (its present value) is determined at the next closest coupon payment date, and the buyer pays the seller the portion of the coupon payment for the time the seller owned the bond. This prorated portion of the coupon payment is referred to as *accrued interest.*

Future Value of an Ordinary Annuity

Figure 4.8 illustrates the calculations required for calculating the future value of an ordinary annuity. Because it is an ordinary annuity, the equal payments, *PMT*, are due at the end of every conversion period for the next n periods. Notice that the payment schedule shown on the x-axis is the same payment schedule as previously shown in Figure 4.7, except that now we are determining the future value, rather than the present value of the payments. Investment plans to which equal contributions are made at the end of each period, and for which a future amount is desired, are typical of this type of time diagram.

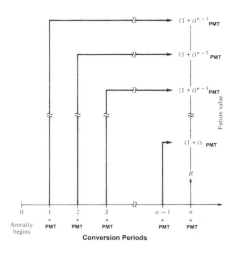

FIGURE 4.8 Future value of an ordinary annuity.

As shown in Figure 4.8, the future value of the payments at the end of the nth period is obtained by finding the future value of each individual payment and then summing the results. Thus,

$$FV = PMT + PMT(1+i) + PMT(1+i)^2 + \ldots + PMT(1+i)^{n-3}$$
$$+ PMT(1+i)^{n-2} + PMT(1+i)^{n-1}$$

Factoring PMT from this equation, we have

$$FV = PMT[1 + (1+i) + (1+i)^2 + \ldots + (1+i)^{n-3} + (1+i)^{n-2} + (1+i)^{n-1}]$$

The terms in brackets are a geometric series, whose sum can be written as:

$$\left[\frac{(1+i)^n - 1}{i} \right]$$

Thus, the Future Value becomes

$$FV = PMT \left[\frac{(1+i)^n - 1}{i} \right] \qquad \text{(Eq. 4.17)}$$

Example 7: Isabel Johnson decides to save for a new iPad by depositing $30 at the end of each month in an account that pays 1.2% interest compounded monthly. How much will she have at the end of six months?

Solution A time diagram for this situation is given in Figure 4.9. The payments are due at the end of each period, hence, this is an example of an ordinary annuity. Observe that because the first payment is not made until the end of the first month, it will draw interest for only 5 months. Similarly, the last payment, made at the end of the 6-month interval, will draw no interest, but will, of course, contribute to the final sum. Using Equation 4.17 with $i = 0.012/12 = 0.01$, $PMT = \$30$, and $n = 6$, we compute

$$FV = \$30 \left[\frac{(1+0.01)^6 - 1}{0.01} \right]$$

$$= \$30 \left[\frac{(1.06152015) - 1}{0.01} \right]$$

$$= \$30 [6.152015]$$

$$= \$184.56$$

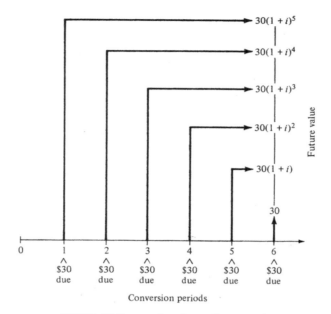

FIGURE 4.9 Future value of an ordinary annuity.

Exercises 4.5

Present and Net Present Value Exercises

1. Determine the present value of a $50 return every quarter for the next 10 years at 4% annual interest compounded quarterly.

2. Determine the present value of an investment that will return $500 at the end of each year for the next 15 years at an annual interest rate of 3% compounded yearly.

3. Determine the NPV of the investment described in Exercise 2 if the annual interest rate is 8% compounded yearly.

4. Ms. Johnson is offered two investment opportunities. The first will return $500 at the end of the year for the next 20 years, while the second will return $1,000 at the end of the year for the next 7 years. Which is more profitable at an annual interest rate of 6%?

5. A woman can invest $70,000 now and receive $2,000 at the end of each month for the next 2 years plus an additional $38,000 at the end of the second year, or she can invest the $70,000 now and receive $3,500 at the end of each month for the next 2 years. Which opportunity is the most profitable at 4% annual interest per annum compounded monthly?

6. Mrs. Wilson has $1,500 to invest. Determine the NPV of her investment if she lends the money to a friend who repays her $750 at the end of each year for the next 3 years. Use an annual interest rate of 3.5%.

7. A man can initially invest $20,000 and receive $2,300 at the end of each quarter for the next 3 years, or he can invest $18,000 now and receive $1,500 at the end of each quarter for the next 4 years. Which opportunity is more profitable at an annual interest rate of 4% compounded quarterly?

8. A bond pays $50 every six months. How much should you pay for the bond (that is, what is its present value) if you want a 4% return on your money and the bond has ten payments left. Additionally, on the tenth payment you will also receive $1,000, which is the original price of the bond.

9. A bond pays $80 every six months. How much should you pay for the bond (that is, what is its present value) if you want a 6% return on your money and the bond has twenty payments left. Additionally, on the twentieth payment you will also receive $1,000, which is the original price of the bond.

Future Value Exercises

10. Determine the future value of an ordinary annuity in which $1,000 is deposited at the end of each year for 10 years at 4% annual interest compounded annually.

11. Determine the future value of an ordinary annuity in which $400 is deposited at the end of each quarter for 3 years at 4% annual interest compounded quarterly.

12. Determine the future value of an ordinary annuity in which $100 is deposited at the end of each month for three years at 2% interest compounded monthly.

13. Mr. Hamedi deposits $20 at the end of each week in an account at 5% annual interest compounded weekly. How much money will he have to spend on holiday gifts when he takes the money out at the end of 48 weeks.

14. To provide for her child's education, a mother deposits $1,000 every June 30 and December 31 for 17 years. Determine the value of the annuity just after her last payment, which occurs on December 31, if her investment plan pays 4% annual interest compounded semiannually.

15. A bond pays $75 every month. How much should you pay for the bond (present value), if you want a 4.25% return on your money and the bond has a 12 payments left. Additionally, on the twelfth payment you will also receive $1,000, which is the original price of the bond.

4.6 MORTGAGES AND AMORTIZATION

One of the most common types of ordinary annuities is a mortgage on a house or land. The mortgage is a loan used to pay for the property, with the property serving as collateral for the loan. This gives the lender, known as the *mortgagor*, a claim on the property should the borrower, known as the *mortgagee*, default on paying the mortgage. Full title to the property is only transferred to the mortgagee when the loan is fully paid.

In a traditional fixed-rate mortgage the monthly payment and interest rate are fixed for the life of the mortgage. Each payment is used to pay both the interest and principal for the loan. First, the monthly interest charge on loan is determined and paid, with the remaining portion of the monthly payment applied to paying off the loan.[7]

Although the monthly payment is fixed, the interest due changes each month, decreasing with every payment. This occurs because the interest is computed each month anew on the unpaid loan balance. As the loan gets paid off, the unpaid balance decreases, which means that the interest due each month also decreases. Thus, each month more and more of the payment gets applied to paying off the loan. This method of payment is commonly referred to as the United States Rule.

The main consideration with mortgages is to determine the amount of the monthly payment, which depends on the original amount of the loan, the interest rate, and the length of the loan. For all mortgages that adhere to the United States Rule, the payment, *PMT*, is determined as

$$PMT = \frac{PV}{\left[\dfrac{1-(1+i)^{-n}}{i}\right]} \qquad \text{(Eq. 4.18)}$$

where:

PMT = the monthly payment
PV = the original amount of the loan
i = the monthly interest rate = (the annual interest rate) / 12
n = the length of the loan, in months, = 12 * (the number of years of the loan)

Notice that Equation 4.18 is the same as Equation 4.15, except that it is used to solve for the value of *PMT* given *PV*, rather than solving for *PV* given the *PMT* amount.

---

[7] With adjustable-rate mortgages, known as ARMs, the interest rate is only fixed for an initial term, but then fluctuates with market interest rates. This provides an initially lower monthly payment than that of a traditional fixed-rate mortgage, but makes the monthly payments unpredictable after the initial term.

Example 1 Mr. Johnson receives a $100,000 mortgage for 30 years at 4% annual interest. Determine his monthly payment.

Solution Twenty years corresponds to $n = (12)(30) = 360$ months, and the monthly interest rate is $i = 0.04/12$. Substituting these values into Equation 4.18, with the initial amount of the loan, $PV = \$100,000$, we find

$$PMT = \frac{\$100,000.}{\left[\dfrac{1-\left(1+\dfrac{0.04}{12}\right)^{-360}}{\left(\dfrac{0.04}{12}\right)}\right]} = \frac{\$100,000.}{\left[\dfrac{1-0.3017959}{\left(\dfrac{0.04}{12}\right)}\right]}$$

$$= \frac{\$100,000.}{\left[\dfrac{.6982041}{\left(\dfrac{0.04}{12}\right)}\right]} = \frac{\$100,000.}{209.46123} = \$477.42$$

Example 2 Mr. Kokowski has agreed to sell a small piece of his property to his neighbor Mr. Brown for $50,000. They agree that Mr. Brown will pay this amount in monthly payments over the next 5 years at 6% interest. How much will each payment be?

Solution Here $PV = \$50,000$, $i = 0.06/12 = 0.005$, and $n = (5)(12) = 60$. Using Equation 4.18, we calculate the monthly payment, PMT, to be

$$PMT = \frac{\$50,000.}{\left[\dfrac{1-\left(1+0.005\right)^{-60}}{\left(0.005\right)}\right]} = \frac{\$50,000.}{\left[\dfrac{1-0.74137220}{\left(0.005\right)}\right]}$$

$$= \frac{\$50,000.}{\left[\dfrac{0.25862780}{0.005}\right]} = \frac{\$50,000.}{51.72556} = \$966.64$$

Determining Total Interest Paid

Once the monthly payment has been determined using Equation 4.18, the total interest paid on the mortgage is easily determined as follows:

$$\textbf{Total Interest Paid} = (\boldsymbol{PMT} * \boldsymbol{n}) - \boldsymbol{PV} \qquad \text{(Eq. 4.19)}$$

Example 3 Determine the total interest paid on a 30 year, 4% mortgage for a loan of $100,000.

Solution The monthly payment for this mortgage was determined to be $477.42 in Example 1. Therefore, using Equation 4.19, the total interest paid when the mortgage is completed is

$$\text{Total Interest Paid} = (\$477.42 * 360) - \$100,000.$$
$$= \$171,871.20 - \$100,000$$
$$= \$71,871.20$$

Example 4 Determine the total interest paid on the mortgage described in Example 2.

Solution The monthly payment for this mortgage was determined to be $966.64. Here $n = 60$ and the $PV = \$50,000$. Using Equation 4.19, the total interest paid for this mortgage is

$$\text{Total Interest Paid} = (\$966.64 * 60) - \$50,000.$$
$$= \$57,998.40 - \$50,000 = \$7,998.40$$

Amortization Schedules

Amortization refers to the repayment of a loan using regular installments over a period of time. Because this is the type of payment used to repay mortgages, mortgage loans are said to *amortized.*

An *amortization schedule* is a table that shows the amount of each payment, and lists the portion of each payment that goes toward paying the interest, the portion of the payment credited against the principal and, finally, the outstanding loan balance after the payment has been made.

Table 4.4 provides an amortization schedule for a loan of $1,000 made at 5% annual interest under the United States Rule. The monthly payments for this loan, which can be verified using Equation 4.18, is $85.61.

In reviewing Table 4.4, first notice that the final outstanding balance at the end of last column is zero, which means the loan has been fully repaid. Next, notice that each payment in the second column is the same monthly payment of $85.61. This is always true, except possibly for the last payment. In general, the last payment is adjusted, if necessary, so that the balance of the loan is zero when the last payment is made. This can occur because of rounding to the nearest cent in the calculations of the monthly payment and interim balances.

TABLE 4.4 Amortization Scheduled for a $1,000 Loan over 12 Months at 5% Interest.

Payment Number	Payment Amount	Interest Paid	Principal Paid	Outstanding Balance
0	–	–	–	$1,000.00
1	$85.61	$4.17	$81.44	$918.56
2	$85.61	$3.83	$81.78	$836.78
3	$85.61	$3.49	$82.12	$754.66
4	$85.61	$3.14	$82.46	$672.20
5	$85.61	$2.80	$82.81	$589.39
6	$85.61	$2.46	$83.15	$506.24
7	$85.61	$2.11	$83.50	$422.74
8	$85.61	$1.76	$83.85	$338.89
9	$85.61	$1.41	$84.20	$254.70
10	$85.61	$1.06	$84.55	$170.15
11	$85.61	$0.71	$84.90	$85.25
12	$85.61	$0.36	$85.25	$0.00

Notice that the interest paid each month (the third column), decreases with each payment, while the principal paid (fourth column) increases with each payment. The values in these two columns are constructed as follows:

The interest due each month is the monthly interest rate times the outstanding balance. Because this is a 5% annual loan, the monthly interest rate, i, is 0.05/12, and the initial outstanding balance is $1,000.

Thus, the interest for the first month is:

$$I_1 = (0.05/12) \, \$1,000 = \$4.17$$

Because part of the monthly payment must be used to cover this interest, we are left with $85.61 − $4.17 = $81.44 as the payment applied to the loan itself. Thus, the outstanding balance at the end of the first month is:

$$P_1 = \$1,000 - \$81.44 = \$918.56$$

For the second month the interest is again calculated as the monthly interest rate times the outstanding balance. Now, however, the outstanding balance is $918.56. Thus,

$$I_2 = (0.05/12) \, \$918.56 = \$3.83$$

and the amount of the payment that is left to be applied to paying the loan balance is $85.61 − $3.83 = 81.78. With this amount applied to the loan, the new outstanding balance at the end of the second month is:

$$P_2 = \$918.56 - \$81.78 = \$836.78$$

Continuing in this manner, the complete amortization schedule listed in Table 4.4 was generated.

Because the outstanding balance is reduced with each payment, the interest owed also declines from payment to payment, and a larger portion of each monthly payment is credited against the loan balance. For larger mortgages having a longer life than the one in this example, the size of the interest portion of each payment, especially in the first few years of the mortgage, is much more dramatic. This is illustrated in the next example.

Example 4 Calculate the first two lines of an amortization schedule for a 30 year, $100,000 loan, at an annual interest rate of 4%.

Solution The monthly payment for this mortgage is $477.22 (see Example 1). Thus, the first month's interest is

$$I_1 = (0.04/12) \, \$100,000 = \$333.33.$$

That is, $333.33 out of the first monthly payment of $477.42 is paid as interest to the institution making the loan (this is 69.82% of the payment). This leaves ($477.42 − $333.33) = 144.09 of the payment to be applied directly to paying off the loan. Thus, the outstanding balance at the end of the first month is

$$P_1 = \$100,000 - \$144.09 = \$99,855.91$$

The interest charge for the second month is again calculated as the monthly interest rate times the outstanding balance. Thus,

$$I_2 = (0.04/12) \, \$99,855.91 = \$332.85$$

and the amount of the payment that is applied to paying the loan balance is $477.42 − $332.85 = 144.57. When this amount applied to the loan, the new outstanding balance at the end of the second month is

$$P_2 = \$99,855.91 - \$144.57 = \$99,711.34$$

Creating Amortization Schedules Using Excel®[8]

Amortization schedules are extremely easy to create using Excel. One additional advantage in using this program is that it provides a payment function for directly calculating the monthly payment. Figure 4.10 shows an Excel spreadsheet for the mortgage described in Example 3.

	A	B	C	D	E	F
1	Amount of Loan:	1000				
2	Length of Loan (in years):	1				
3	Annual Interest Rate:	5%				
4	Monthly Payment:	$85.61				
5						
6		Payment Number	Payment Amount	Interest Paid	Principal Paid	Outstanding Balance
7		0				$1,000.00
8		1	$85.61	$4.17	$81.44	$918.56
9		2	$85.61	$3.83	$81.78	$836.78
10		3	$85.61	$3.49	$82.12	$754.66
11		4	$85.61	$3.14	$82.46	$672.20
12		5	$85.61	$2.80	$82.81	$589.39
13		6	$85.61	$2.46	$83.15	$506.24
14		7	$85.61	$2.11	$83.50	$422.74
15		8	$85.61	$1.76	$83.85	$338.89
16		9	$85.61	$1.41	$84.20	$254.70
17		10	$85.61	$1.06	$84.55	$170.15
18		11	$85.61	$0.71	$84.90	$85.25
19		12	$85.61	$0.36	$85.25	$0.00

FIGURE 4.10 A sample amortization table.

The relevant formulas used to create the spreadsheet shown in Figure 4.10 are shown in Figure 4.11. In particular, look at the formula entered into cell B4, which uses the PMT() function. This function computes the mortgage payment that we have been calculating by hand using Equation 4.19.

The PMT() function requires three parameters; the monthly interest rate, the length of the loan, in months, and the amount of the loan. For the spreadsheet shown in Figure 4.11 the PMT() entry in cell B4 is

$$= -PMT(B3/12, B2*12, B1)$$

---

[8] Excel is a registered trademark of Microsoft Corporation.

The reason for the negative sign in front of the PMT() function has to do with the convention of money flows. Positive values, such as the amount of the loan, are considered funds that flow *to the* borrower. Negative values indicate funds that the borrower pays; as such, they are considered funds that *flow away* from the borrower. Thus, the PMT() function will report a negative amount as the monthly payment, because this is an amount that effectively "flows away" from the borrower. To counteract this, the negative sign in front of the PMT() function forces the calculated payment to appear as a positive value.

Note also that once the formulas have been placed in Row 8, they can be copied down to the end of the spreadsheet, as shown by the arrowed lines in Figure 4.11.

	A	B	C	D	E	F	G
1	Amount of Loan:	1000					
2	Length of Loan (in years):	1					
3	Annual Interest Rate:	0.05					
4	Monthly Payment:	=-PMT(B3/12,B2*12,B1)					
5							
6			Payment Number	Payment Amount	Interest Paid	Principal Paid	Outstanding Balance
7			0				=B1
8			1	=$B$4	=$B$3/12*G7	=D8-E8	=G7-F8
9			2	=$B$4	=$B$3/12*G8	=D9-E9	=G8-F9
10			3	=$B$4	=$B$3/12*G9	=D10-E10	=G9-F10
11			4	=$B$4	=$B$3/12*G10	=D11-E11	=G10-F11
12			5	=$B$4	=$B$3/12*G11	=D12-E12	=G11-F12
13			6	=$B$4	=$B$3/12*G12	=D13-E13	=G12-F13
14			7	=$B$4	=$B$3/12*G13	=D14-E14	=G13-F14
15			8	=$B$4	=$B$3/12*G14	=D15-E15	=G14-F15
16			9	=$B$4	=$B$3/12*G15	=D16-E16	=G15-F16
17			10	=$B$4	=$B$3/12*G16	=D17-E17	=G16-F17
18			11	=$B$4	=$B$3/12*G17	=D18-E18	=G17-F18
19			12	=$B$4	=$B$3/12*G18	=D19-E19	=G18-F19

FIGURE 4.11 Formulas for Figure 4.10's amortization table.

Exercises 4.6

1. Determine the monthly payment for a 30 year, $36,000 mortgage, having a 4% interest rate.

2. Determine the monthly payment for a 25 year, $30,000 mortgage, having an 8% interest rate.

3. Determine the total interest paid for the mortgage in Exercise 1.

4. Determine the total interest paid for the mortgage in Exercise 2.

5. Either using a calculator or Excel, determine the sum of the individual interest payments in column D of Figure 4.10 and then verify that this same sum is obtained using Equation 4.19 for this loan.

6. Mr. O'Toole agrees to sell his business to Mr. Johnson. The mortgage obtained by Mr. Johnson for this business is for $20,000 over 5 years at 5%.

    a. Determine the monthly payments.

    b. Determine the total interest paid for this mortgage.

    c. Complete the first three lines of an amortization schedule for this mortgage.

7. Ms. Tilson agrees to sell some property to a friend. The mortgage obtained by the friend for this property is for $45,000 over 4 years at 4%.

    a. Determine the monthly installment.

    b. Determine the total interest paid for this mortgage.

    c. Complete the first three lines of an amortization schedule for this mortgage.

8. Determine the monthly payment for a 25 year, $40,000 mortgage, having a 7.5% interest rate.

9. Determine the total interest paid for the mortgage in Problem 8.

10. Create an Excel spreadsheet that produces an amortization schedule for an $800 loan amortized over 1 year with monthly payments at 4% interest.

11. Create an Excel spreadsheet that calculates the payment and produces the amortization schedule for the mortgage in Exercise 1.

4.7 INSTALLMENT LOANS AND INTEREST CHARGES

Unlike mortgages, which are ordinary annuities that calculate monthly interest based on the unpaid balance of the loan, some commercial loans, such as vacation loans, home improvement loans, and a host of other cash advances for specific purposes, can use two related but different interest determination methods. These are known as the *add-on method* and *discount method*, respectively.

Add-On Installment Loans

In the add-on method of interest and payment calculations, the total finance charge for a loan is determined as

$$\textbf{Total Finance Charge}$$
$$\textbf{= (Annual Rate) (Amount of Loan) (Length of the loan, in years)} \quad \text{(Eq. 4.20)}$$

The monthly installment payment, PMT, is then obtained by adding the total finance charge to the amount of the loan and then dividing this result by the total number of months, n, in the life of the loan.

$$PMT = \frac{(\textit{Total Finance Charge}) + (\textit{Amount of the loan})}{(\textit{Length of the loan, in months})} \quad \text{(Eq. 4.21)}$$

Example 1 A $1,000 loan is negotiated for 2 years at 8% annual interest under the add-on method. Determine the monthly payment.

Solution Using Equation 4.20, the total finance charge is

$$\text{Total Finance Charge} = (.08)(\$1,000)(2) = \$160.$$

Because it is a 2 year loan, the length of the loan, in months, is 24. Therefore, the monthly installment payment, PMT, is obtained using Equation 4.21, as

$$PMT = (\$160 + \$1,000) / 24 = \$1,160 / 24 = \$48.33$$

If the borrower in this example thinks he is paying 8% interest for his loan, he is badly mistaken. In fact, he is paying a good deal more. To determine the actual interest, we can use Equation 4.15, which is repeated below, for convenience

$$PV = PMT \left[\frac{1 - (1+i)^{-n}}{i} \right]$$

Now, however, because PV, PMT, and n are known, and we are trying to solve for i, we will rewrite the equation as

$$\left[\frac{1 - (1+i)^{-n}}{i} \right] = \frac{PV}{PMT}$$

Unfortunately, this is a non-linear equation that is not easily solved for i except by numerical techniques that are beyond the scope of this text. However, most financial calculators that compute mortgage payments using Equation 4.19 can also be used to solve for interest rates. Using such a calculator, the interest rate corresponding to a payment of $48.33, for a 24-month loan of $1,000 reveals a true annual interest rate of 14.69%. This is a good deal higher than the quoted rate of 8%.

The discrepancy between the quoted rate and the actual annual rate is due to the borrower not having full use of the loan for its entire duration. Even though the borrower has control of the full amount of the loan only for the first month, they are charged interest as if they had the full amount for the entire life of the loan. To counteract this discrepancy, all loans now require that the true annual percentage rate, or APR, be specified for all commercial consumer loans (calculation of this true annual rate is presented in Section 4.9)

Example 2 Determine the monthly installment payment for a $22,000 loan for 5 years at 4% using the add-on method. Using a financial calculator, determine the true annual interest rate for this loan.

Solution To calculate the installment payment, PMT, we must first determine the total interest charge. Using Equation 4.20, this charge is

$$\text{Total Finance Charge} = (.04)(\$22{,}000)(5) = \$4{,}400.$$

Because it is a 5-year loan, the length of the loan, in months, is 60. Therefore, the monthly installment payment, PMT, obtained using Equation 4.21, is

$$PMT = (\$4{,}400 + \$22{,}000) / 60 = \$26{,}400 / 60 = \$440.00.$$

Using a financial calculator, with a payment of $440, a present value of $22,000, which is the amount received by the borrower, and a loan length of 60 months yields a true annual interest rate of 7.42%, which is almost twice the stated rate.

Discount Installment Loans

In a discount installment loan, the installment payment, PMT is first determined by dividing the amount of the loan by the number of months in the life of the loan. Mathematically, this can be expressed as

PMT = (Amount of the Loan) / Length of loan, in months) (Eq. 4.22)

Next, the total interest charge is calculated exactly as in the add-on method using Equation 4.20, repeated here for convenience as Equation 4.23.

$$\text{Total Finance Charge}$$
$$= (\text{Annual Rate})(\text{Amount of Loan})(\text{Length of the loan, in years}) \quad \text{(Eq. 4.23)}$$

Finally, the total interest charge is subtracted from the face value of the loan, and the difference is the cash received by the borrower.

$$\text{Cash Received} = \text{Amount of Loan} - \text{Total Interest Charge}$$
$$\text{Amount of Loan}\,[1 - (\text{Annual Rate})(\text{Length of the loan, in years})]$$

—or—

$$\text{Cash Received} = \text{Amount of Loan}$$
$$- \text{Amount of Loan}\,(\text{Annual Rate})(\text{Length of the loan, in years})]$$

—or—

$$\text{Cash Received} =$$
$$\text{Amount of Loan}\,[1 - (\text{Annual Rate})(\text{Length of the loan, in years})] \quad \text{(Eq. 4.24)}$$

Example 3 Determine the monthly installment payment for a $12,000, 5-year loan discounted at 4%. Additionally, determine the total interest charged for this loan and the cash received from the loan.

Solution. Because this is a 5 year loan, the length of the loan, in months, is 60. Therefore, the monthly installment payment, *PMT*, for the loan, using Equation 4.22 is

$$PMT = (\$12{,}000)/60 = \$200.00$$

Using Equation 4.23 the total interest charge is

$$\text{Total Interest Charge} = (.04)(\$12{,}000)(5) = \$2{,}400.$$

The cash received when the loan is made is given by Equation 4.26 as

$$\text{Cash Received} = \$12{,}000 - \$2{,}400 = \$9{,}600.$$

Effectively, the borrower has obtained a $9,600 loan and is paying interest on a $12,000 loan. Thus, as in the add-on method, the quoted rate in the discount method is not comparable to a true stated annual rate. The annual interest rate can be determined using a financial calculator with a payment of $200, a present value of $9,600, which is the cash amount received by the borrower, and a loan length of 60 months. This yields a true interest rate of 9.15% rate for Example 3.

Because the interest charge is subtracted from the original amount of the loan, if a desired cash amount is needed, the borrower must request a higher original amount when the loan request is made. Equation 4.24 can be algebraically rewritten to determine the amount of the loan requested to provide a specific cash amount. Letting r denote the annual interest rate, t, the term of the loan in years, and PV, the cash received by the borrower yields

$$Amount\ of\ Loan = PV = \frac{Cash\ Received\ by\ borrower}{1 - rt} \qquad \text{(Eq. 4.25)}$$

Similarly Equation 4.23 can be rewritten as:

Total Interest Charge =
    Amount of Loan – Cash Received by Borrower   (Eq. 4.26)

Example 4 Bill Bagly is considering a 4-year loan to provide him with $10,000. The financial institution providing the loan has a current loan rate of 5% and uses the discount method for calculating interest. How much should Bill request as the amount of the loan.

Solution Here, the desired cash received is $10,000, $t = 4$, and $r = 0.05$. Using Equation 4.25, the amount of the loan requested should be

Amount of Loan = $10,000 /[1 –(0.05)(4)] = $10,000 / [1 – 0.2]
= $10,000 / 0.8 = $12,500

The total interest charge for this loan, using Equation 4.26, is

Total Interest Charge = $12,500 – $10,000 = $2,500

Because the original loan is for $12,500, and the total interest charge of $2,500 is subtracted from this loan when it is made, Bill will receive $10,000 in cash.

Exercises 4.7

1. Mr. Johnson borrows $3000 from his bank for a new car loan at 8% interest add-on. Determine the monthly installment and the total interest paid if the loan is to be repaid over.

    a. 2 years.

    b. 3 years.

    c. If you have a financial calculator, determine the actual, annual rate for each of these loans.

2. A financial institution advertises automobile loans for 2%.

    a. Determine the total interest paid for a $8,200 loan for 3 years, if the loan is made using the add-on method.

    b. If you have a financial calculator, determine the actual annual rate for this loan.

3. The Bakers have decided to apply for a one year, $4,000, vacation loan that is calculated using the add-on method with a 6% interest rate.

    a. Determine the total interest paid for this loan.

    b. Determine the monthly payment for this loan.

    c. If you have a financial calculator, determine the actual, annual rate for this loan.

4. Assume that the loan described in Exercise 3 is a discount loan rather than an add-on loan.

    a. Determine the monthly installment.

    b. Determine the amount of money the Bakers will receive when the loan is granted.

    c. Using a financial calculator, determine the actual, annual rate for this loan.

5. a. Determine the total interest charge for this loan.

    b. Determine the actual cash received when this loan is granted.

    c. Using a financial calculator, determine the true annual rate for this loan.

6. a. Determine the total interest charge for this loan.

    b. Determine the actual cash received when this loan is granted.

    c. If you have a financial calculator, determine the true annual rate for this loan.

7. Mark needs $15,000 to purchase a car, and takes out a 5 year, discounted loan at a 4% interest rate.

    a. Determine how much must he borrow to realize $15,000 after the loan is granted.

    b. Determine the monthly payment for the loan amount determined in part a.

    c. Determine the total interest charge for the loan amount determined in part a.

    d. If you have a financial calculator, determine the true annual rate for the loan determined in part a.

8. Ms. Smith's application for a $25,000 home improvement discounted loan is approved by a lender for a 5% annual discount loan for 4 years.

    a. Determine his monthly payments.

    b. Determine the actual cash he has available for improvements when the loan is granted.

    c. Determine the total interest paid on this loan.

    d. How much should Ms. Smith apply for if she wants to realize $25,000 when the loan is approved?

    e. Using a financial calculator, determine the true annual rate for the $25,000 loan.

4.8 ANNUITIES DUE

An *annuity due*, which is also referred to as an *immediate annuity*, is one in which the periodic payments and/or receipts begin immediately. An example of this is a lottery payout in which the winner receives annual payments beginning immediately. As with ordinary annuities, simple formulas exist for determining both the present and future values of immediate annuities.

The Present Value of an Annuity Due

Figure 4.12 illustrates a typical ordinary annuity, with *PMT* dollars due at the start of each conversion period for the next *n* periods and where the present value of the complete cash flow is desired.

As shown in Figure 4.12, the present value of an immediate annuity is given by:

$$PV = PMT + PMT(1 + i)^{-1} + PMT(1 + i)^{-2} + \ldots + PMT(1 + i)^{-(n-1)} \quad \text{(Eq. 4.27)}$$

Notice that this equation is essentially the same equation as that for the present value of an ordinary annuity, with the addition of one initial immediate payment at the beginning and one less payment at the end of the equation. Equation 4.27 can be factored as:

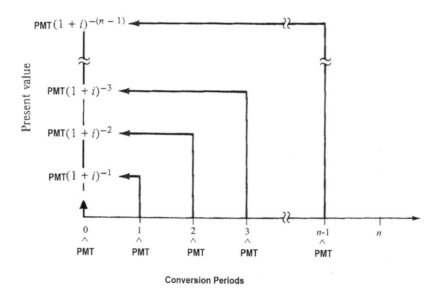

FIGURE 4.12 Present value of an annuity due.

$$PV = PMT + PMT[(1+i)^{-1} + (1+i)^{-2} + (1+i)^{-3} + ... + (1+i)^{-(n-1)}] \qquad \text{(Eq. 4.28)}$$

The terms in brackets represent the present value of an ordinary annuity with $n-1$ payments remaining. The terms in brackets are a geometric series, whose sum can be written as:

$$PMT[(1+i)^{-1} + (1+i)^{-2} + (1+i)^{-3} + ... + (1+i)^{-(n-1)}] = \left[\frac{1-(1+i)^{-(n-1)}}{i} \right]$$

Thus, the Present Value becomes

$$PV = PMT + PMT(1+i) \left[\frac{1-(1+i)^{-n}}{i} \right] \qquad \text{(Eq. 4.29)}$$

Equation 4.25 can be algebraically manipulated, as shown below, to yield a second formula for determining an annuity due's present value

$$PV = PMT + PMT\left[\frac{1-\left(1+i\right)^{-(n-1)}}{i}\right]$$

$$= PMT\left[\frac{i+1-\left(1+i\right)^{-(n-1)}}{i}\right]$$

$$= PMT\left[\frac{\left(1+i\right)-\left(1+i\right)^{-(n-1)}}{i}\right]$$

$$= PMT\left(1+i\right)\left[\frac{1-\left(1+i\right)^{-n}}{i}\right]$$

Thus,

$$\boldsymbol{PV = PMT\left(1+i\right)\left[\frac{1-\left(1+i\right)^{-n}}{i}\right]} \qquad \text{(Eq. 4.29a)}$$

As both versions, Equation 4.29 and 4.29a yield the same result, use whichever formula you prefer.

Example 1 Oliver Bronson wins the New Jersey million dollar lottery. Instead of taking a cash settlement, he decides to take 20 equal annual payments of $50,000, starting immediately. Determine the present value of this annuity if the current interest rate is 3%.

Solution This is an example of an annuity due with $n = 20$, $i = 0.03$, and $PMT = \$50,000$. Substituting these values into Equation 4.29, we obtain

$$PV = \$50,000 + \$50,000\left[\frac{1-\left(1+0.03\right)^{-19}}{0.03}\right]$$

$$= \$50,000 + \$50,000\left[\frac{1-0.57028603}{0.03}\right]$$

$$= \$50,000 + \$50,000\left[\frac{0.42971397}{0.03}\right]$$

$$= \$50,000 + \$50,000\left[14.323799\right] = \$766,189.95$$

Notice that if the single cash payment is greater than this amount, it is preferable to take the single cash payment, because this cash payment can be invested at 3% and provide a higher dollar return over time. If the single cash payment is less than the present value of the annuity, the annuity payments are the preferable option.

The Future Value of an Annuity Due

Figure 4.13 illustrates the calculation of the future value for an annuity due in which PMT dollars is paid at the beginning of every conversion period for the next n periods.

For the annuity shown in Figure 4.13, the future value of the equal payment cash flows at the end of the nth period is:

$$FV = PMT(1 + i) + PMT(1 + i)^2 + \ldots + PMT(1+ i)^{n-2} + PMT(1+ i)^{n-1} + PMT(1+ i)^n$$

Factoring $PMT(1 + i)$, from this equation yields

$$FV = PMT\,(1 + i)[1 + (1 + i) + \ldots + (1+ i)^{n-3} + (1+ i)^{n-2} + (1+ i)^{n-1}] \quad \text{(Eq. 4.30)}$$

The terms in brackets in Equation 4.30 are a geometric series, whose sum can be written as:

$$\left[\frac{(1+i)^n - 1}{i} \right]$$

Thus, the Future Value becomes

$$\boldsymbol{FV = PMT\,(1+i)} \left[\frac{(1+i)^n - 1}{i} \right] \quad \text{(Eq. 4.31)}$$

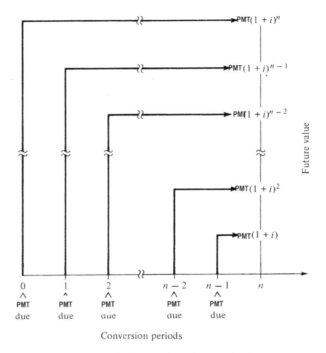

FIGURE 4.13 Future value of an annuity due.

Equation 4.27 can be algebraically manipulated, as shown below, to yield a second formula for determining an annuity due's future value

$$FV = PMT\left(1+i\right)\left[\frac{\left(1+i\right)^{n}-1}{i}\right]$$

$$= \frac{PMT\left(1+i\right)^{n+1} - PMT - PMT}{i}$$

$$= \frac{PMT - PMT}{i} - \frac{PMT\,i}{i}$$

$$\left[= PMT\frac{\left(1+i\right)^{n+1}-1}{i}\right] - PMT$$

Thus,

$$\boldsymbol{FV = PMT}\left[\frac{\left(\boldsymbol{1+i}\right)^{\boldsymbol{n+1}}-\boldsymbol{1}}{\boldsymbol{i}}\right] - \boldsymbol{PMT} \qquad \text{(Eq. 4.31a)}$$

As both versions, Equations 4.31 and 4.31a yield the same result; use whichever formula you prefer.

Example 2 The Jacksons decide to save $100 every half-year, beginning immediately, for home improvements that they anticipate making in 4 years. How much will they have at the end of the fourth year if they make eight deposits in an account yielding 4% interest compounded semiannually.

Solution A time diagram for this situation is given in Figure 4.14. Because the payments are due at the beginning of each time period, this is an example of an annuity due. In particular, the first payment draws interest for all eight periods, and the last payment, made at the beginning of the eighth period, also earns interest. Here, then, $PMT = 100$, $n = 8$, and $i = 0.04/2 = 0.02$. Substituting these values into Equation 4.31, we obtain

$$FV = \$100(1+0.02)\left[\frac{(1+0.02)^8 - 1}{0.02}\right] = \$100(1.02)\left[\frac{(1.17165938) - 1}{0.02}\right]$$

$$= \$100(1.02)[8.582969] = \$875.46$$

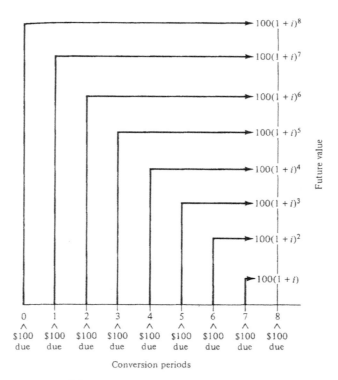

FIGURE 4.14 Future value of an annuity due.

Example 3 Determine the future value after 11 years of an annuity due of $45 per quarter at 4% per annum compounded quarterly.

Solution Here $PMT = 45$, $n = 44$, and $i = 0.04/4 = 0.01$

$$FV = \$45(1+0.01)\left[\frac{(1+0.01)^{44} - 1}{0.01}\right]$$

$$= \$45(1.01)\left[\frac{(1.54931757) - 1}{0.01}\right]$$

$$= \$45(1.01)[54.931757] = \$2,496.65$$

Exercises 4.8

Present and Net Present Value Exercises

1. Determine the present value of an annuity due in which $1000 is deposited at the beginning of each year for 10 years at 4% annual interest compounded annually.

2. Determine the present value of an annuity due in which $40 is deposited at the beginning of each quarter for 3 years at 4% annual interest compounded quarterly.

3. Determine the present value of an annuity due in which $20 is deposited at the beginning of each month for 15 years at 2% compounded monthly.

4. Determine the present value of an annuity due in which $100 is deposited on January 1 and July 1 for 12 years at 3% compounded semiannually.

Future Value Exercises

5. Determine the future value of an annuity due in which $40 is deposited at the beginning of each quarter for 3 years at 4% annual interest compounded quarterly.

6. Determine the future value of an annuity due in which $1000 is deposited at the beginning of each year for 10 years at 4% annual interest compounded annually.

7. Mr. Gouzien deposits $20 at the beginning of each week in an account at 5% annual interest compounded weekly. How much money will he have to spend on holiday gifts when he takes the money out at the end of 48 weeks.

8. Determine the future value of an annuity due in which $100 is deposited at the beginning of each month for 10 years at 2% compounded monthly.

9. Determine the total value after 2 years of an annuity due of $1 per day at 5% annual interest compounded daily.

10. To provide for his child's education, Mr. Gouzein deposits $1,000 every January 1st and July 1st for 16 years. Determine the value of the annuity at the end of the 16th year, which is six months after the last payment. Assume his investment plan pays 4% annual interest compounded semiannually.

11. In 2014, Andy Gregg decided to invest $1,000 at the beginning of each year into a no-load mutual fund. Determine what the value of his holding will be at the end of 2020, if the fund's stock increases in value at 8% per year.

4.9   ANNUAL PERCENTAGE YIELD (APY) AND RATE (APR)

In practice, interest rates are stated in a number of different ways, some of which reflect compounding and some of which do not. The most common of the terms used in providing interest rates are listed in Table 4.5.

TABLE 4.5 Standard Interest Rate Measures.

Term	Meaning	Notation
Nominal Interest Rate also referred to as the **Nominal Annual Interest Rate** and the **Stated Annual Interest Rate**	The interest rate expressed on a per-year basis. It is a simple interest rate that does not take into account how many times a year the interest rate is compounded.	r
Effective Interest Rate also referred to as the **Annual Percentage Yield**	The annual interest rate that takes into account how many times a year interest is compounded.	E_f and APY
Annual Percentage Rate	The annual interest rate that does not take into account compounding.	APR

Nominal Interest Rate

The nominal interest rate, which is frequently referred to as the nominal rate, for short, is a simple interest that does not take into account any compounding. For example, a nominal interest rate of 4% means that every dollar deposited will earn 4% interest, which is $4, at the end of one year.

Clearly, for example, interest compounded quarterly will generate more interest than the same amount compounded annually. In particular, if $100 is invested at a nominal rate of 8% compounded quarterly, the balance at the end of the year, which consists of 4 quarters, is $100 $(1 + .08/4)^4$ = $108.24. That is 8% compounded quarterly is equivalent to an annual nominal interest rate of 8.24%. Here, the 8% is the annual rate and the 8.24% is referred to as the effective interest rate. That is, it will require an effective interest rate of 8.24% to generate the same interest income as a nominal rate of 8% compounded quarterly.

Effective Interest Rate

The effective interest rate, which is also referred to as both the Effective Rate and the Annual Percentage Yield, or APY for short, is an annual rate that generates the same interest as a compounded nominal rate. For example, as seen above, a nominal rate of 8% compounded quarterly has an effective interest rate of 8.24%. Thus, the effective interest rate takes into account the compounding of the nominal annual rate.

Effective rates, denoted as E_f are determined using the formula

$$E_f = \left(1 + r/n\right)^n - 1 \qquad \text{(Eq. 4.32)}$$

where:

r is the nominal annual rate, and

n *is* the total number of compound periods per year for the nominal annual rate.

Example 1 Determine the effective rate, E_f for an interest-bearing account with a nominal annual rate of 12% compounded.

a. quarterly,

b. monthly,

c. weekly, and

d. daily.

Solution

a. $E_f = (1 + 0.12 / 4)^4 - 1 = 1.125509 - 1 = 0.125509$ or 12.55%.

Thus, 12% compounded quarterly is equivalent to a nominal annual rate of 12.55%.

b. $E_f = (1 + 0.12/12)^{12} - 1 = 1.126825 - 1 = 0.126825$ or 12.68%.

Thus, 12% compounded monthly is equivalent to a nominal annual rate of 12.68%.

c. $E_f = (1 + 0.12/52)^{52} - 1 = 1.127359 - 1 = 0.127359$ or 12.74%.

Thus, 12% compounded weekly is equivalent to a nominal annual rate of 12.74%.

d. $E_f = (1 + 0.12/365)^{365} - 1 = 1.127475 - 1 = 0.127475$ or 12.75%.

Thus, 12% compounded daily is equivalent to a nominal annual rate of 12.75%.

Notice from this example that the effective rate increases as n, the number of conversion periods in a year, increases.

The effective interest rate, as already noted, is also referred to as the annualized percentage rate, or APY, for short. It is sometimes is also referred to as the effective annual percentage rate, or EAPR, for short. More commonly, however, the annual percentage rate, or APR, is a nominal rate as described next.

Annual Percentage Rate (APR)

The annual percentage rate is a nominal annual rate determined as follows:

$$Nominal\ APR = (Interest\ rate\ per\ compounding\ period)$$
$$(Compounding\ periods\ in\ a\ year) \qquad (Eq.\ 4.33)$$

Example 2 A credit card company advertises that the interest rate on its credit card has a 29% APR with a daily compounding rate of .07942%. Determine if this stated APR is correct?.

Solution Using Equation 4.33, the APR associated with a daily interest rate of .079452% is:

$(.079452\%)\ (365) = $ is 29%.
Thus, the stated APR is correct.

It should be noted that nominal APR does not take into account compounding. By simply multiplying the interest rate per compounding period by the number of compounding periods, as defined in Equation 4.33, it effectively reproduces the same periodic interest rate n times, which is the definition of a simple interest rate.

Example 3 Determine the effective interest rate for the 29% APR given in Example 2.

Solution Using Equation 4.32, the effective annual rate, E_f, is

$$E_f = (1 + 0.0007945212/365)^{365} - 1 = 1.33627 - 1 = 0.33627 \text{ or } 33.63\%.$$

Continuous Compounding

The largest effective interest that can be generated from a stated interest rate, i, is when it is compounded continuously. This continuous compounding rate, is:

$$E_c = e^i - 1$$

which is referred to as the *continuous conversion rate*, where $e = 2.71828\ldots$ is Euler's number.

Example 4 Determine the continuous conversion rate for a stated annual rate of 12%.

Solution $E_c = e^{0.12} - 1 = (2.71828)^{0.12} - 1 = 1.127497 - 1 = 0.127497$ or 12.7497%, which is a rate not too different from that achieved by daily compounding.

The formula required to determine the dollar amount of interest earned at a continuous rate is:

$$Continuous \ Interest = P \ e^{it} \qquad \text{(Eq. 4.34)}$$

where

> P is the principal
> i is the nominal annual rate, compounded continuously
> t is the time, in years

Example 5 Determine the interest on $1000 when it is deposited in an account for three years that pays 8% compounded continuously.

Solution. Using Equation 4.34, with $P = \$1,000$, $e = .08$, and $t = 3$ yields:

$$Interest = \$1000 \ e^{(.08)3} = \$1000 \ (1.271249) = \$ \ 127125$$

Exercises 4.9

In Exercises 1 through 8, determine the effective interest rate for the following annual nominal rates.

1. 2% compounded quarterly.

2. 2% compounded daily (assume 365 days in a year)

3. 2% compounded continuously.

4. 4% compounded semiannually.

5. 4% compounded continuously.

6. 8% compounded annually.

7. 8% compounded quarterly.

8. 8% compounded continuously.

4.10 SUMMARY OF KEY POINTS

Key Terms

- Amortization
- Amortization schedule
- APR
- Annuity
- Annuity due
- Compound interest
- Conversion period
- Duration
- Effective interest rate
- Future value of an annuity
- Future value of a lump sum
- FV
- i
- Interest
- Interest rate per period
- Maturity
- n
- N

- Nominal interest rate
- Ordinary annuity
- Payment interval
- Present value of an annuity
- Present value of a lump sum
- Principal
- PMT
- PV
- Regular payment
- Simple annuity
- Simple interest

Key Concepts

Interest

- With simple interest, interest payments are identical from one time period to the next because the interest is always calculated on the original investment or deposit.
- With compound interest, interest payment are made on the current balance of an account, which is generally different from the original balance.
- With compound interest, interest payments on a lump-sum deposit increase from one conversion period to the next.

Present Value

- A dollar today is worth more than a dollar due at any time in the future.
- Different investment opportunities can be compared only when the value of each is known at the same instant in time. To compare competing opportunities that mature at different times, compare either their present values or the value of each opportunity at the same future instant of time.
- To calculate the present value of an investment with multiple payments, determine the present value of each individual payment and then sum the results.

Annuities

- An annuity consists of equal payments made at successive and equal periods of time.
- An annuity is simple if the compounding period at which interest is calculated coincides with the payment period.

- An ordinary annuity is an annuity in which all payments are made at the end of each payment period.
- A traditional mortgage is an example of an ordinary annuity.
- An annuity due is an annuity in which all payments are made at the beginning of each payment period.

Present Value of an Annuity

- The present value of an annuity is the amount that must be deposited at the beginning of the annuity to yield the desired payments to be made by the annuity. Thus, it is the present value of all annuity payments.
- The fair market value of an annuity is its present value. Because a bond is an ordinary annuity, the price of a bond is the present value of all of its payments plus the present value of what will be collected at maturity.

RATES OF CHANGE: THE DERIVATIVE

In this Chapter

In business, rates of change are used to determine whether sales, revenue, profit, and other measures of a company or economy have been growing or declining. Additionally, a rate of change puts a numerical value on how strong or weak this growth or decline has been. It can also provide an indication of what may lie in the future, assuming the conditions that produced the change continue.

This chapter begins by presenting the average rate of change between two quantities, a topic that is typically familiar to most students. The central focus of the chapter, however, is the concept of an instantaneous rate of change. This rate of change, which is known as a derivative, provides the rate of change at a single point, rather than between two points.

The derivative has many applications in finance, economics, operations research, inventory control, physics, and other business and scientific areas. Table 5.1 provides a very short list of how three of these areas refer to the derivative.

TABLE 5.1 Usage of the Derivative in Finance, Economics and Physics.

Profession	Terminology	What it is
Finance	Bond Duration	First derivative of bond price function
Finance	Bond Convexity	Second derivative of bond price function
Economics	Marginal Revenue	First derivative of revenue function
Economics	Marginal Cost	First derivative of cost function
Physics	Velocity	First Derivative of distance function
Physics	Acceleration	Second derivative of distance function

In addition to its use in finance, economics, the physical and social sciences, derivatives are also used to solve a large class of optimization problems. We reserve our study of these applications, however, until Chapter 6. In this chapter, we direct our attention to the derivative as a rate of change and providing specific rules for calculating derivatives of commonly encountered mathematical functions.

5.1 AVERAGE RATES OF CHANGE

An extremely useful measure in business forecasting is the average rate at which a quantity changes. For example, it is useful in predicting sales for the month of April to know that sales in March totaled 10,000 units. Even more useful, however, is the additional information that sales have been increasing, on average, at a rate of 2,000 units per month. Similarly, a company needs information on wage scales for the past year before it can prepare next year's budget. In addition, however, information on the average yearly increase in salaries is also important. In both these situations knowledge of the rates that quantities change, either changes in sales per month or changes in wages paid per hour, increases one's ability to forecast future requirements accurately.

One rate of change you are probably familiar with is average speed, which measures the change in distance for a given change in time. For example, if an individual traveled 500 miles in 10 hours, the average change in distance with respect to time is 50 miles per hour. That is, on the average, each hourly increase in driving time resulted in an increase of 50 miles traveled. Formally, this average speed is calculated by dividing the total change in miles driven by the total change in hours traveled using the formula:

$$Average\ speed = \frac{Change\ in\ distance}{Change\ in\ time}$$

Example 1: A woman drove from her house to a destination 400 miles away. The following table shows how far she traveled at 1 hour intervals. Determine her average speed for

a. the entire trip

b. the last four hours of her trip

Time (in hours)	0	1	2	3	4	5	6	7	8
Distance (in miles)	0	50	115	175	230	280	290	350	400

Solution:

a. Average speed for the trip $= \dfrac{Change\ in\ distance}{Change\ in\ time} = \dfrac{400 - 0}{8 - 0} = \dfrac{400\ miles}{8\ hours} = 50\ mph$

b. Average speed for last 4 hours $= \dfrac{Change\ in\ distance}{Change\ in\ time} = \dfrac{400 - 230}{8 - 4} = \dfrac{170\ miles}{4\ hours}$

$$= 42.0\ mph$$

As seen in this example, the average speed can vary as the distance and time interval over which we measure them changes. In each case, however, the average rate of change is calculated by dividing the total change in miles driven by the total change in hours traveled.

In its most general form, an average rate of change in one quantity with respect to a second quantity is defined as the ratio of two changes. Denoting the first and second quantities by y and x, respectively, we have

$$Average\ rate\ of\ change\ in\ y\ with\ respect\ to\ x = \frac{change\ in\ y}{change\ in\ x}. \qquad (Eq.\ 5.1)$$

Example 2 Using the information presented in Figure 5.1, find the average rate of change in sales with respect to months over the 3-month period, April through June.

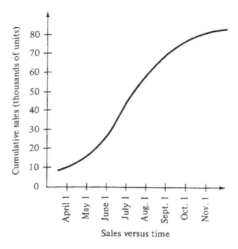

FIGURE 5.1 Monthly sales.

Solution Reading directly from the graph, we find that total sales as of April 1 were 10,000 units, while total sales at the end of June (July 1) were 44,000. The change in sales is 44,000 − 10,000 = 34,000, and the change in time is 3 months. Therefore,

$$\text{Average rate of change} = \frac{34{,}000}{3} = 11{,}333 \text{ units per month}$$

Example 3 Redo Example 2 for the 3-month period, June through August.

Solution Reading directly from Figure 5.1, we find that total sales as of June 1 were 26,000, while total sales at the end of August (September 1) were 69,000.

The change in sales is 69,000 − 26,000 = 43,000 units, and the change in time is 3 months. Therefore,

$$\text{Average rate of change} = \frac{43{,}000 \text{ units}}{3 \text{ mo}} = 14{,}333 \text{ units per month}$$

Examples 2 and 3 together again illustrate the point that average rates of changes depend on the intervals under consideration. Different intervals typically will result in different average rates of change.

In computing rates of change, the tacit assumption is that the first quantity, in some manner, is related to the the second quantity. In Example 1, we assumed that distance travelled was a function of time (the more time spent traveling, the greater the distances covered) and that in Example 2, that recording sales in time is meaningful.

When dealing with mathematical functions, where y is a known function of x, that is, $y = f(x)$, we can rewrite Equation 5.1 somewhat more neatly. Suppose we are interested in the average rate of change over the interval $[x_1, x_2]$ where the notation $[x_1, x_2]$ denotes all values of x between x_1 and x_2 inclusive. The change in x is simply $x_2 - x_1$. At x_1, the associated value of y is $y_1 = f(x_1)$, whereas at x_2 the associated value of y is $y_2 = f(x_2)$, The change in y is $y_2 - y_1$ or, equivalently, $f(x_2) - f(x_1)$. Therefore, Equation 5.1 can be written as:

$$\text{Average rate of change in } y \text{ with respect to } x = \frac{f(x_2) - f(x_1)}{x_2 - x_1}. \qquad \text{(Eq. 5.2)}$$

Example 4 Determine the average rate of change in the function illustrated in Figure 5.2 between the points $x = 1$ and $x = 4$.

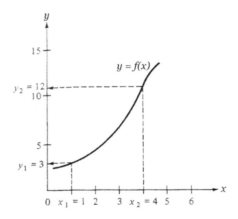

FIGURE 5.2 Determining an average rate of change.

Solution At the points $x_1 = 1$ and $x_2 = 4$ the corresponding y-values are $y_1 = 3$ and $y_2 = 12$. Thus, over the 3-unit interval from $x_1 = 1$ to $x_2 = 4$, the value of the function (y-values) changes 9 units. Using Equation 5.1 we obtain

$$\text{Average rate of change} = \frac{12 - 3}{4 - 1} = \frac{9 \text{ units of } y}{3 \text{ units of } x} = 3 \text{ units of } y \text{ per unit of } x.$$

If the relationship between y and x is given by an algebraic equation rather than a graph, we can use Equation 5.2 to calculate the average rate of change directly.

Example 5 Determine the average rates of change in the function $f(x) = x^2 - 2x + 4$ over the interval $x = 2$ to $x = 5$, and then over the interval $x = 2$ to $x = 4$.

Solution For $x_1 = 2, f(x_1) = f(2) = 2^2 - 2(2) + 4 = 4$ and, for $x_2 = 5, f(x_1) = f(5) = 5^2 - 2(5) + 4 = 19$. Using Equation 5.2, we compute

$$\text{Average rate of change} = \frac{f(5) - f(2)}{5 - 2} = \frac{19 - 4}{5 - 2} = 5.$$

Thus, on the average, over the interval $x = 2$ to $x = 5, f(x)$ changes 5 units for every 1-unit change in x.

Over the interval $[2, 4]$, we have

$$\text{Average rate of change} = \frac{f(x_2) - f(x_1)}{x_2 - x_1} = \frac{f(4) - f(2)}{4 - 2} = \frac{12 - 4}{4 - 2} = 4.$$

Thus, over the interval $x = 2$ to $x = 4$, the function changes four units on average for each oneunit change in x.

The right side of Equation 5.2 should be familiar; it was used in Chapter 2 to define the slope of a straight line. Because the physical significance of the slope is a rate of change, our work in this section reinforces our previous results. It also extends these results to all functions. For any function $y = f(x)$, whether it is a linear equation or not, the quantity $\dfrac{f(x_2) - f(x_1)}{x_2 - x_1}$ represents an average rate of change.

Geometric Significance

In addition to a value, average rates of change also have geometric significance. When $y_1 = f(x_1)$ and $y_2 = f(x_2)$, the average rate of change over the interval $[x_1, x_2]$ is, in fact, **the slope of the straight line** through the points (x_1, y_1) and (x_2, y_2). This relationship is illustrated in Figure 5.3.

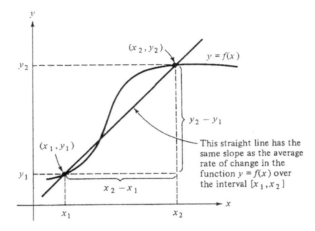

FIGURE 5.3 An average rate of change is a slope.

Returning to Example 2, we see that we can calculate the average rate of change for April through June by first drawing a straight line through the two points in Figure 5.6 corresponding to April 1 and July 1. This is done in Figure 5.4. The slope of this line is the average rate of change.

FIGURE 5.4 The average rate of change as a slope.

To graphically determine the average rate of change for the months of June through August, we first draw a straight line through the points in Figure

5.5 corresponding to June 1 and September 1. This is done in Figure 5.5. The slope of this line is the average rate of change over the months June through August. Note that its slope is different from the one in Figure 5.4.

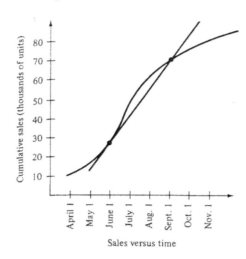

FIGURE 5.5 Graphically determining an average rate of change.

Exercises 5.1

In Exercises 1 through 10, find the average rate of change for each function over the given interval.

1. $f(x) = x^2 - 4x + 5$ over the interval $[1, 8]$

2. $f(x) = x^2 - 4x + 5$ over the interval $[-2, 1]$

3. $f(x) = 4x^2 - 6$ over the interval $[1, 3]$

4. $f(x) = 1/x$ over the interval $[1, 4]$

5. $f(x) = x^2 + 3x - 1$ over the interval $[2, 6]$

6. $f(x) = e^x$ over the interval $[0, 2]$

7. $f(x) = 3x - 4$ over the interval $[1, 5]$

8. $f(x) = x^2 + 6x + 2$ over the interval $[1, 5]$

9. $f(x) = x^3 + 5$ over the interval $[1, 5]$

10. $f(x) = 2 - 3t^2$ over the interval $[1, 5]$.

5.2 INSTANTANEOUS RATES OF CHANGE

Although average rates of change are useful for many decision-making purposes, they are not always sufficient. Sometimes events change so rapidly that weekly, daily, and even hourly average rates of change are not current enough. Sometimes information is needed to determine how quickly something is changing *right now*. Rather than an average rate of change that measures change over an interval, this requires an *instantaneous rate of change*, which measures change at an instant of time.

An approximation of the instantaneous rate of change at a single input value of x, which we will label as x_1, can be obtained by taking the average rate of change over smaller and smaller intervals, always having x_1 as a fixed endpoint. This process is clearer when viewed graphically. Figure 5.6 depicts the average rate of change in a function $y = f(x)$ over the interval $[1, 4]$.

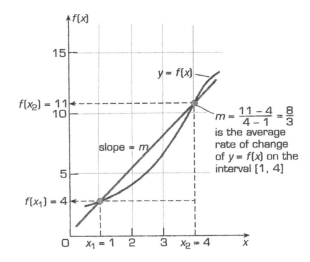

FIGURE 5.6 The average rate of change over the interval $[4, f(4)]$.

To approximate the instantaneous rate of change of the function shown in Figure 5.6 at the value $x_1 = 1$, we find the average rates of change of $f(x)$ over smaller and smaller intervals, with each interval having $x_1 = 1$ as an endpoint. This approach is shown in Figure 5.7, which is a graph of the same function $f(x)$ shown in Figure 5.6. The approximation is found by taking the average rate of change of $f(x)$ over smaller and smaller intervals, each with x_1 as its endpoint.

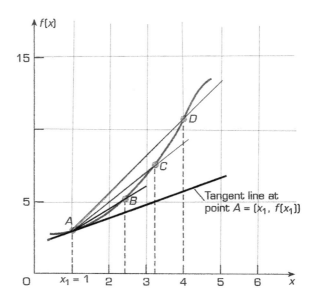

FIGURE 5.7 The instantaneous rate of change as the limit of the average rate of change.

In Figure 5.7, the straight line between the points A and D represents the average rate of change of the function over the interval $[1, 4]$, the line between A and C represents the average rate of change of the function over the interval $[1, 3.2]$, and finally, the straight line between A and B represents the average rate of change over the interval $[1, 2.4]$. Each of these lines is formally known as a *secant line*. Notice as the intervals get smaller, the slopes of the secant lines approach the slope of the darker line that just touches point A. *It is the slope of this darker line that is defined as the instantaneous rate of change of the function $f(x)$ at the point $x_1 = 1$.*

The darker line shown in Figure 5.7 is formally known as a *tangent line*.[1] It is the slope of the tangent line at a point on a curve that is defined as the function's instantaneous rate of change at that point. For each point that has a tangent line then, it has a defined instantaneous rate of change.

Two important observations are relevant here. The first is that a tangent line may not exist at all points on a curve. When this is the case, the instantaneous rate of change is not defined for the function at these points. Cases, where this occurs, are presented at the end of the next section.

---

[1] A tangent line is defined as a straight line that locally touches a curve at one and only one point A tangent line may touch or cross the curve at other points, as long as each of these other point is, at least, slightly away from the first point.

The second observation is that the instantaneous rate of change can, and almost always does, differ at different points of a function. Thus, because there is an infinity of points on a curve, and graphically determining the tangent line's slope for even a few of these points becomes unrealistic. For example, consider Figure 5.8, where two tangent lines are shown – one at $x = -1.3$ and one at $x = 2$. It is the slope of each of these lines that is the instantaneous rate of change of the function at the points shown.

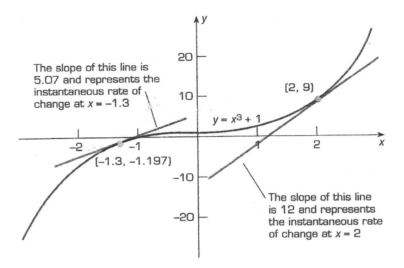

FIGURE 5.8 Instantaneous rates of change may vary at different input values along the curve.

Fortunately, there is a mathematical procedure for easily determining slopes of all tangent lines, where they exist, at all points on the curve. The basis for this procedure is as follows.

As previously shown in Figure 5.7, as the intervals for which the average rates of change are calculated get smaller and smaller, the slopes of the secant lines for these intervals approach the slope of the tangent line. To obtain a mathematical formula for this procedure, let point A in Figure 5.7 has coordinates (x_1, y_1) and point D coordinates (x_1, y_1), as shown in Figure 5.9.

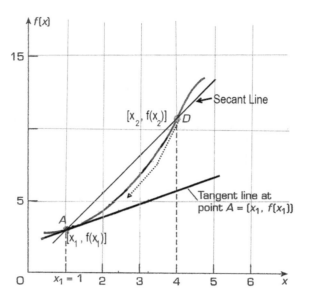

FIGURE 5.9 Secant lines converging onto a tangent line.

The slope of the secant line through the two points A and D, whose coordinates are (x_1, y_1) and (x_2, y_2), respectively, where $y_1 = f(x_1)$ and $y_2 = f(x_2)$, is given by (see Equation 5.2)

$$\frac{f(x_2) - f(x_1)}{x_2 - x_1}$$

In Figure 5.9, the limit of this slope for the secant line between the points A and D, as the D point moves along the curve closer to A, is the slope of the tangent line at the point A. Thus, as previously noted, the slope of the tangent line, when it exists, is the limit of the slopes of the secant lines. Mathematically, this is expressed as follows:

$$\textit{The slope of the tangent line at } x_1 = \lim_{x_2 \to x_1} \frac{f(x_1) - f(x_2)}{x_2 - x_1} \qquad \text{(Eq. 5.3)}$$

Where $\lim_{x_2 \to x_1}$ is read "the limit as x_2 approaches x_1."

The concept of one value approaching another value, in this case x_2 approaching x_1 is fundamental to all of calculus. We are interested in knowing what happens as x_2 gets arbitrarily close to x_1; the words *arbitrarily* and *close* are both important. Thus, *we are not interested* in what happens when x_2 equals x_1, but only in the result as x_2 gets close to x_1. Just how close is another matter, and we use the word *arbitrarily* to signify as close as the mind can conceive."

When a tangent line exists at a point (x_1, y_1) on the graph of the function $y = f(x)$, the limit defined in Equation 5.3 will exist at that point. Because this slope is the value of the instantaneous rate of change at the given point, Equation 5.3 can be rewritten as:

The Instantaneous Rate of Change in f(x) at x_1

$$= \lim_{x_2 \to x_1} \frac{f(x_1) - f(x_2)}{x_2 - x_1} \qquad \text{(Eq. 5.4)}$$

In the next section, we show that the calculus term *derivative* is another name for the instantaneous rate of change. That is, the derivative of a function at a point is the instantaneous rate of change of the function at that point. We then show how to use Equation 5.4 to find the derivative (the instantaneous rate of change) of a function $f(x)$ for any value of x, for which the limit in Equation 5.4 exists.

Units

Instantaneous rates of change, as their average rates of change counterparts, measure the change in a dependent variable versus a change in its corresponding independent variable. As such, the units associated with both of these rates of change are the same.

For example, the average rate of change calculated in Example 1 of the previous section was the change in distance versus a change in time, more commonly known as speed. This measure is the same for both the average over a long trip or the instantaneous measure provided by a car's speedometer.

Example 1 A function that measures fuel efficiency for automobiles gives the distance traveled, m, for the amount of gasoline used, in gallons, g. That is, $m = f(g)$. Determine the units associated with the instantaneous rate of change of f with respect to gasoline consumption.

Solution m is measured in units of miles and g is measured in units of gallons. Thus, the instantaneous rate of change of m with respect to g has units of mile per gallon, commonly known as *mpg*.

Example 2 The graph in Figure 5.10 shows the cost of producing a textbook as a function of the number of copies produced. Determine the units associated with the instantaneous rate of change of this function with respect to the size of a production run.

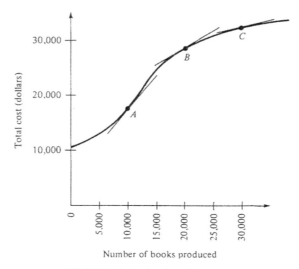

FIGURE 5.10 Textbook production costs.

Solution The independent variable, x, is the number of books produced (a decision made by the publisher), and the dependent variable, y, is the total cost, in dollars, associated with production. The instantaneous rate of change is in units of dollars per book.

The curve shown in Figure 5.10 is the graph of a cost function; in this case, the cost as a function of books produced. There are three tangents drawn at points A, B, and C, resopectively, corresponding to production runs of 10,000, 20,000, and 30,000, respectively. The slope of each tangent line, in units of dollars per book, is a marginal cost. The slope of the tangent line at A is greater than the slope of the tangent line at B, which in turn is greater than the slope of the tangent line at C. Consequently, the marginal cost at a production run of 10,000 books is greater than the marginal cost of a production of 20,000 books, which is greater than the marginal cost at a production run of 30,000 books. That is, in dollars/book, the additional cost of producing book 10,001 is greater than the additional cost of producing book 20,001, which in turn is greater than the additional cost of producing book 30,001. As more books are produced, it becomes cheaper per book to produce additional books.

All of Chapter 6 is devoted to discussing other important commercial applications of instantaneous rates of change. Here we are simply introducing the terms to suggest the relevance of instantaneous rates of change to commerce and economics.

Exercises 5.2

In Exercises 1 through 4, use the tangent line that appears in each figure to estimate the instantaneous rate of change of each function at the x-component of the point of tangency.

1.

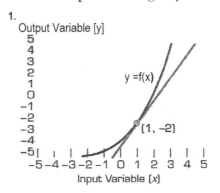

2.

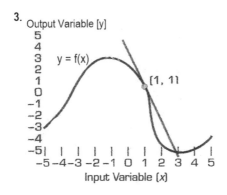

3.

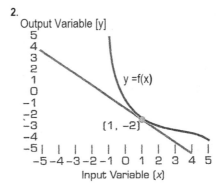

4.

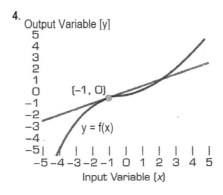

5. Sketch the function $f(x) = x^2 - 6x + 10$ and graphically determine the slope of the tangent lines at the points:

   a. $(1, 5)$ **b.** $(3, 1)$ **c.** $(5, 5)$.

6. Sketch the function $f(x) = x^2 - 4x + 5$ and graphically determine the slope of the tangent line at $(3, 2)$.

7. Sketch the function $f(x) = x^2 + x$ and graphically determine the slope at $x = 2$.

8. Consider the function

$$y = \begin{cases} 2 + 5x & (x \le 3) \\ 23 - 2x & (x > 3) \end{cases}$$

which is sketched in Figure 5.11.

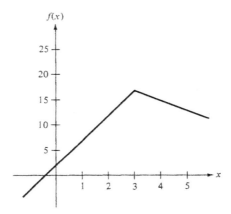

FIGURE 5.11

a. Determine graphically whether this curve has a tangent at $x = 3$.

b. Compute average rates of change for this function over the intervals having $x = 3$ as one endpoint and the second endpoint given successively as 3.2, 2.8, 3.1, 2.9, 3.05, 2.95, 3.01, 2.99, 3.005, and 2.995. Do these rates approach a fixed value? What can you conclude about the instantaneous rate of change at $x = 3$?

9. Consider the function

$$y = \begin{cases} 7 & (x \le 5) \\ 12 & (x > 5) \end{cases}$$

which is sketched in Figure 5.12. Calculate average rates of change for this function over the intervals having $x = 5$ as one endpoint and the second endpoint given successively by 3, 7, 4, 6, 5.3, 5.3, 5.7, 5.3, 5.9, 5.1, 5.95, 5.05, 5.99, and 5.01. What is the instantaneous rate of change in this function at $x = 5$?

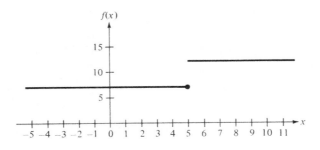

FIGURE 5.12

10. A company's sales are known to be related to advertising expenditures by the equation $S(x) = 150{,}000 + 6000x - 50x^2$, where x denotes the monthly advertising expenditures in thousands of dollars.

   a. Find the instantaneous rate of change in sales with respect to advertising expenditures.

   b. Using your answer to part a, determine whether an increase in advertising would increase sales if the present advertising budget is $45,000. Would the situation be different with a$60,000 advertising budget?

11. A firm's total sales, in millions of dollars, are given by the function $R(x) = 3x + \frac{1}{2}x^2$, where x denotes the number of years the firm has been in operation.

   a. Determine the firm's average growth rate in sales for its first 7 years in business.

   b. Determine the firm's instantaneous rate of growth after its seventh year in business.

   c. What will the firm's total sales be at the end of the tenth year if sales continue to follow the given equation?

   d. What will the firm's total sales be at the end of the tenth year if the growth in sales after the seventh year always equals the growth achieved at the end of the seventh year?

12. Determine the values of x for the function illustrated in Figure 5.13 between or at which the instantaneous rate of change in y with respect to x is:

   a. positive,

   b. negative,

   c. zero.

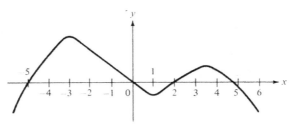

FIGURE 5.13

13. Determine the values of x for the function illustrated in Figure 5.14 between or at which

   a. the derivative is positive,

   b. the derivative is negative, and

   c. the derivative is zero.

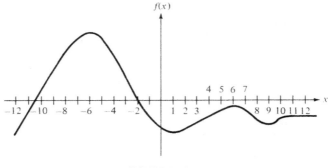

FIGURE 5.14

14. From past experience, it is known that an increase in the price of wheat leads to a decrease in the demand for wheat. Based on this information, determine which of the following equations may possibly relate the demand, D, for wheat to its price, p.

   a. $D = 30p^2 + 5p + 2000$

   b. $D = 3000/p$

   c. $D = 5p^2 - 2000$

   d. $D = 5p^2 - 2000p$.

15. From past experience it is known that an increase in the price of the silver leads to an increase in the demand for this commodity. Based on this information, determine which of the following equations may possibly relate to the demand, D, for silver to its price, p.

a. $D = \dfrac{200}{p}$

b. $D = 300p^2 + 50 + 200$

c. $D = 2000 - 2p^3$

c. $D = \dfrac{150}{p^2}$

5.3 THE DERIVATIVE

As noted in the previous section, the instantaneous rate of change in a function is also known as the *derivative* of the function. Thus, the following definition is an immediate consequence of the equivalence of these two terms:

Definition 5.1 If the limit

$$= \lim_{x_2 \to x_1} \frac{f(x_1) - f(x_2)}{x_2 - x_1} \qquad \text{(Eq. 5.5)}$$

exists, this limit is called the derivative of the function $y = f(x)$ at the point x.

Rather than working with Equation 5.5 directly, it is sometimes easier first to change notation by setting $h = x_2 - x_1$. Here h represents the distance between x_2 and x_1. On solving, $h = x_2 - x_1$ for x_2, we also have $x_2 = x_1 + h$, and can replace the requirement "x_2 approaches x_1" with the requirement "h approaches 0." That is, because $x_2 = x_1 + h$, the condition that h approaches 0 is equivalent to the condition that x_2 approaches x_1. Substituting the quantities $h = x_2 - x_1$, $x_2 = x_1 + h$, and $\lim_{x_2 \to x_1} = \lim_{h \to 0}$ into Equation 5.5, we obtain

The derivative of the function $f(x)$ *at* x_1 $= \lim_{h \to 0} \dfrac{f(x_1 + h) - f(x_1)}{h}$ (Eq. 5.6)

where this limit exists.

Equation 5.6 gives the derivative (i.e., the instantaneous rate of change) of a function $y = f(x)$ at a specific point x_1. Generally, one needs the derivative at many points. For example, in the textbook 'case illustrated in Figure 5.10, we found the instantaneous rate of change, that is, the derivative, at 'three points. To find the derivative, we can proceed in one of two ways.

The first way is to evaluate Equation 5.6 separately at each point x_1 that we need. The second, and more efficient procedure is to evaluate Equation 5.6 at an arbitrary point $x_1 = x$. Doing this yields an equation in terms of the variable x that can then evaluated at any desired point, x_1 by simply substituting the value of x_1 into the result. This is done as follows:

Replacing x_1 by x in Equation 5.6, we conclude that the

$$The\ derivative\ of\ f(x)\ at\ any\ point\ x = \lim_{h \to 0} \frac{f(x+h) - f(x)}{h} \qquad \text{(Eq. 5.7)}$$

Equation 5.7 can now be used to derive a general expression for the derivative of $f(x)$ at any point x. As just noted, once we have this general expression, the derivative at any specific point is found by evaluating the expression at the required point.

The following four-step process can be used to determine the derivative using Equation 5.7.

Step 1 Find $f(x + h)$ by replacing x with the quantity $x + h$ in the given function $f(x)$.

Step 2 Calculate $f(x + h) - f(x)$ by subtracting $f(x)$ from the expression for $f(x + h)$ found in Step 1. Simplify the resulting difference.

Step 3 Divide by h.

Step 4 Let h approaches zero.

Example 1 Determine a general expression for the derivative of the function $f(x) = x^2$ using Equation 5.7. Then, determine the derivative when:

a. $x = 1$,

b. $x = 3$,

c. $x = 5$.

Solution Using the recommended four-step procedure for evaluating Equation 5.7 with $f(x) = x^2$, we have

Step 1: $f(x + h) = (x + h)^2$,

Step 2: $f(x + h) - f(x) = (x + h)^2 - x^2$
$$= (x^2 + 2xh + h^2) - x^2$$
$$= 2xh + h^2$$

Step 3: $\dfrac{f(x+h) - f(x)}{h} = \dfrac{2xh + h^2}{h} = \dfrac{h(2x + h)}{h} = 2x + h$

Step 4: As h approaches zero, the expression $2x + h$ approaches $2x$. Therefore,

$$\lim_{h \to 0} \frac{f(x+h) - f(x)}{h} = \lim_{h \to 0} (2x + h) = 2x.$$

Thus, the derivative of the function $f(x) = x^2$ at any point x *is* $2x$. Evaluation this derivative at the desired points yields:

a. when $x = 1$, the value of the derivative is $2(1) = 2$.

b. when $x = 3$, the value of the derivative is $2(3) = 6$

c. when $x = 5$, the value of the derivative is $2(5) = 10$

Example 2 Determine a general expression for the derivative of the function $f(x) = 2x^2 + 3x - 2$.

Solution Following the four-step procedure, we have

Step 1: $f(x + h) = 2(x + h)^2 + 3(x + h) - 2$

Step 2: $f(x + h) - f(x) = [2(x + h)^2 + 3(x + h) - 2] - [2x^2 + 3x - 2]$
$$= [2x^2 + 4xh + 2h^2 + 3x + 3h - 2] - [2x^2 + 3x - 2]$$
$$= 4xh + 2h^2 + 3h$$

Step 3: $\dfrac{f(x+h) - f(x)}{h} = \dfrac{4xh + 2h^2 + 3h}{h} = \dfrac{h(4x + 2h + 3)}{h} = 4x + 2h + 3$

Step 4: As h approaches zero, the expression $4x + 2h + 3$ approaches $4x + 3$. Therefore,

$$\lim_{h \to 0} \frac{f(x+h) - f(x)}{h} = \lim_{h \to 0} (4x + 2h + 3) = 4x + 3.$$

which is a general expression for the value of the derivative in $f(x) = 2x^2 + 3x - 2$ at any point x.

The four-step procedure used in both previous examples to determine a derivative can be always be used to find the derivative of any function when the derivative exists. Its real value, however, is that it allows the development of more general rules that are simpler and much quicker for calculating derivatives. These rules are the subject of the next four sections.

We conclude this section with a few observations. First, the general expression for the derivative of a given function at an arbitrary point x *is itself a function* of x. In Example 1, we found the derivative of $f(x) = x^2$ to be $2x$, while in Example 2 we found the derivative of $f(x) = 2x^2 + 3x - 2$ to be $4x + 3$.

This is not at all surprising. Because most functions are not linear equations, it follows that the graphs of such functions are not straight lines. Therefore, most curves have different tangents at different points on the curve, each tangent line with its own slope. As the slopes themselves are dependent on where they are taken, they are also functions of x.

Secondly, throughout this section we have assumed that x_2 was always greater than x_1 or, equivalently, that h was always positive. This need not be the case. As we take smaller and smaller intervals, we require only that x_1 always be in each interval. It need not be the left-hand endpoint. If a derivative exists the answer obtained for the value of the derivative using x_1 as the 'left endpoint of each interval is the same answer obtained using x_1 as the right–hand endpoint. Only if both approaches yield the same answer can we mathematically say that a value of the derivative exists. For most functions encountered in commercial application this is the case Where the two limits differ, a single limit does not exist, and, subsequently, there is no derivative. The existence of these points is presented next.

Existence of a Derivative

Because derivatives are defined as limits (see Eq. 5.7), wherever a point lacks a limit, a derivative does not exist there. Alternatively, because the limit at a point is also the slope of the tangent line at that point, this can be restated as: "if there is no tangent line at a point, there is no derivative at that point." Commonly occurring cases where tangent lines and, hence, derivatives do not exist are the following:

First, tangent lines, and, thus, derivatives do not exist at sharp corner points. Such a point is shown in Figure 5.15 at $x = 1$. At this point, a tangent line cannot be drawn

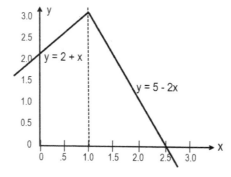

FIGURE 5.15 Derivatives do not exist at corner points.

A second place where derivatives do not exist and a single tangent line cannot be drawn is where a function has a hole or a ledge, such as that shown in Figure 5.16. At such points, the function is defined as *discontinuous*. A function is defined as *continuous* at all points where it is seamless, and its graph has no holes or ledges. Derivatives exist wherever a function is continuous and do not exist at any points of discontinuity. The formal definition of continuity is presented at the end of this section.

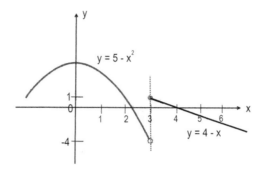

FIGURE 5.16 A Derivative does not exist where there is a Hole or Ledge.

Because the graph shown in Figure 5.17 has a discontinuity at $x = 3$ and is continuous (i.e., seamless) everywhere else, it has no derivative at $x = 3$ and the derivative exists at all other points on the graph.

The final case where derivatives do not exist is where a function either is not defined or a tangent line is vertical. These cases are illustrated in Figure 5.17.

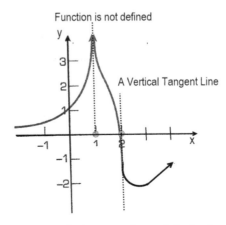

FIGURE 5.17 Two further cases where a derivative does not exist.

Derivatives and Continuity

Both Figures 5.15 and 5.16 are graphs of *piecewise* functions (review Section 2.2 if you are unfamiliar with this type of function). However, Figure 5.16 contains a gap, also referred to as a ledge, at the point $x = 3$, while Figure 5.15 has no such gaps. Where a gap, hole, or ledge exists in a curve, the curve is said to have a *discontinuiuty*. Thus, Figure 5.16 has a discontinuity at $x = 3$, while Figure 5.15 has no discontinuities.

Mathematically, Figure 5.15 is said to be continuous at all points, while Figure 5.16 is continuous at all points *except* at $x = 3$, where it is discontinuous. Formally a function $f(x)$ is defined as continuous at a point $x = a$ if the following three conditions are satisfied[2]:

1. $f(a)$ is defined

2. $\lim\limits_{x \to \bar{a}} f(x) = f(a) \leftarrow$ this is called a left – hand limit.

3. $\lim\limits_{x \to a^+} f(x) = f(a) \leftarrow$ this is called a right – hand limit.

where the notation

$\lim\limits_{x \to a^-} f(x)$ is read as "the limit of $f(x)$ as x approaches the value $x = a$ from the left"

And

$\lim\limits_{x \to a^+} f(x)$ is read as " the limit of $f(x)$ as x approaches the value $x = a$ from the

right,"[3]

Condition (1) requires that there be a well-defined functional value at the value $x = a$. Conditions (2) and (3) require that this functional value be the same one when approaching $x = a$ from both the left and from the right sides; that is approaching the point a from values of $x < a$ and then approaching $x = a$ from values of $x > a$. If any of these three conditions do not hold, the function is discontinuous at the point a.

---

[2] This definition *does not apply* to endpoints because endpoints are only defined at one side. Thus, either a right-side or left- side limit will not exist at an endpoint. Endpoint continuity is presented at the end of this section.

[3] Note that a left-hand limit means the point is being approached from the left along the function and **not** that it is on the left side of a line. Similarly, a right-hand limit means the point is approached from the right and **not** that it is on the right side of a line.

Example 3 Is the function defined by the equations

$$y = f(x) = \begin{cases} 5 - x^2 & x < 3 \\ 4 - x & x \geq 3 \end{cases}$$

continuous when $x = 3$.

Solution For this function to be continuous, the three conditions defining continuity must be satisfied.

At the point $x = 3$, $f(3)$ is clearly defined and has the value 1. This is because the equation defining $f(x)$ at $x = 3$ is given by $4 - x$ for $x \geq 3$. Thus, $f(3) = 4 - 3 = 1$.

For the second condition,

$$\lim_{x \to 3^-} f(x) = f(1)$$

to be satisfied, this same functional value must be the limit as we approach $x = 3$ from the left. Approaching from the left means, we are coming into $x = 3$ from values to its left; that is from values less than 3. For this reason, we must use the equation $f(x) = 5 - x^2$ that defines the function when x is less than 3. Therefore we write

$$\lim_{x \to 3^-} (5 - x^2) = 5 - 9 = -4.$$

Because the left-hand limit is not $f(3) = 1$, the second condition is not satisfied. This means that the function is not continuous when $x = 3$. Another way of saying this is that the function is discontinuous at $x = 3$.[4]

It should be noted that when approaching $x = 3$ from the right, and using the equation that defines the function from this side ($x \geq 3$), we have

$$\lim_{x \to 3^+} (4 - x) = 4 - 3 = 1$$

which satisfies the third condition.

If a function is continuous at a point $x = a$, there will be a seamless transition of functional values as x moves from near a onto a itself and then away from a, from whichever direction the point is approached. This provides a simple and intuitive test for continuity at a point; namely that a function is continuous at all points where its graph has no gaps, holes or ledges. At any

---

[4] It should be noted that when approaching $x = 3$ from the right, and using the equation that defines the function from this side ($x \geq 3$), we have $\lim_{x \to 3^+} (4 - x) = 4 - 3 = 1$, which satisfies the third condition.

point where a gap, hole or ledge exists, that is, where you would have to pick up your pencil when drawing the graph, the function is discontinuous.

Example 4 Graphically determine if the function defined in Example 3 is continuous when $x = 3$. The graph of this function is reproduced as Figure 5.18 with added notation for determining left and right-side limits.

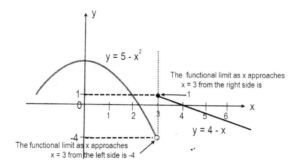

FIGURE 5.18 Approaching a point from the left and right.

Solution Because there is a gap in the curve at $x = 3$ the function whose graph is shown in Figure 5.18 is discontinuous at $x = 3$.

Wherever a derivative exists at a point defined by a function, the function is continuous at that point; however, a function can be continuous at a point and not have a derivative at that point. For example, the function previously shown in Figure 5.15 and reproduced as Figure 5.19, with added notations, is continuous at all points but does not have a derivative when $x = 1$. Thus, although the existence of a derivative at a point ensures that the function is continuous at that point, a function may be continuous at a point and not have a derivative there.

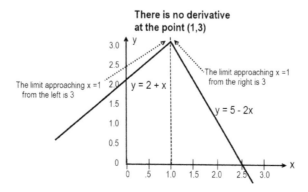

FIGURE 5.19 Continuity at point does not ensure a derivative.

Endpoints

Because endpoints are only defined on one side of a line, either a left-side or right-side limit may exist for such points, but not both. As such, continuity for endpoints is defined as either continuous from the right or the left.

For a function to be continuous from the right at a point *a* the following two conditions must hold:

1. $f(a)$ is defined

2. $\lim_{x \to \bar{a}} f(x) = f(a) \leftarrow$ the limit coming in from the right must equal $f(a)$

For a function to be continuous from left at a point *a* the following two conditions must hold:

1. $f(a)$ is defined

2. $\lim_{x \to \bar{a}} f(x) = f(a) \leftarrow$ the limit coming in from the right must equal $f(a)$

It is worth noting that endpoint continuity is defined *from the direction from which the endpoint is approached, not on the side that the endpoint is located* on the function. Thus, an endpoint located on the left side of a line can only be continuous from the right because it can only be approached from the right. Similarly, an endpoint located on the right side of a line can only be continuous from the left because it can only be approached from the left.

Example 5 Determine if either of the two endpoints shown in Figure 5.20 are continuous from either the left or right.

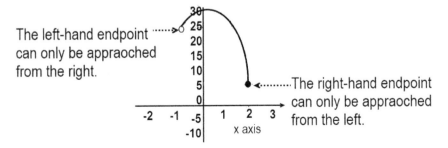

FIGURE 5.20 A graph with two endpoints.

Solution Consider the left-hand endpoint first.

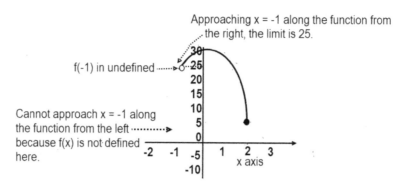

FIGURE 5.21 The left-hand endpoint is not continuous from the right.

As shown in Figure 5.21, the limit as $x = -1$ is approached exists and is equal to 25. However, the function is not continuous from the right at this point because $f(-1)$ is not defined. (Recall from Section 2.2 that an open circle means there is no functional value at the circle point.)

Now consider the right-hand endpoint. As shown in Figure 5.22, the limit approaching $x = 2$ from the left is 5. Because $f(2)$ is defined as 5 (recall that a closed circle means there is a functional value at the circle), and the limit of the function when approached from the left equals this value, the function is continuous from the left.

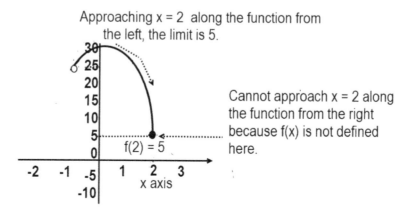

FIGURE 5.22 The right-hand endpoint is continuous from the left.

Exercises 5.3

In Exercises 1 through 10, use Equation 5.7 and the four-step procedure presented in this section to determine the derivatives of the following functions. Then use each derivative to find the instantaneous rate of change at $x = 1$ and $x = 5$.

1. $f(x) = 2 - 3x$

2. $f(x) = 3x - 2$.

3. $f(x) = x^2 - 2x + 10$

4. $f(x) = \frac{1}{2}x^2 - 2x + 10$

5. $f(x) = 2x^3 - 2x^2 + 3x - 1$

6. $f(x) = x^3$

7. $f(x) = \dfrac{x}{x + 2}$

9. $f(x) = \dfrac{1}{x}$

8. $f(x) = \dfrac{1}{x^2}$

10. $f(x) = \dfrac{1}{x} + x$

For each of the functions defined in Exercise 11 through 14, determine limits from the right and from the left as $x = 1$, $x = 2$, and $x = 3$. Then decide if the functions are continuous at these points.

11.

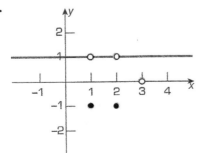

12.

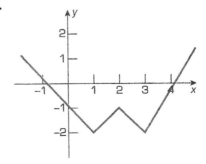

13.

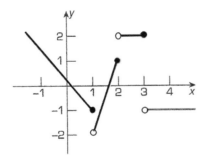

14.

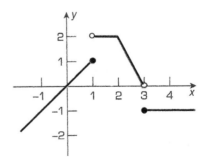

5.4   BASIC DERIVATIVES

As presented in the previous section, the derivative of the function $y = f(x)$ at any point x is

$$\lim_{h \to 0} \frac{f(x+h) - f(x)}{h}$$

The process of finding a derivative is called *differentiation*. Four common notations for the derivative of a function are:

$f'(x) \leftarrow$ read as "f prime of x"

$y' \leftarrow$ read as "y prime"

$\dfrac{dy}{dx} \leftarrow$ read as "dee y dee x"

$\dfrac{d}{dx}[f(x)] \leftarrow$ read as "dee f of x dee x"

Note that dy/dx does *not* mean dy divided by dx, and $f'(x)$ does *not* mean f' times x.

The choice of notation is one of personal preference or convenience; we use them interchangeably in this text. Each set of symbols are only notations for "the derivative of $f(x)$ with respect to the independent variable." Thus, using the notation $f'(x)$, we have

$$f'(x) = \lim_{h \to 0} \frac{f(x+h) - f(x)}{h}$$

As the derivative of the *function $f(x)$* with respect to x, denoted as $f'(x)$, it is also the instantaneous rate of change of $f(x)$ at the point $(x, f(x))$, as well as the slope of the tangent line to the graph of $f(x)$ at the point $(x, f(x))$.

In Example 1 of the previous section we found that the derivative of the function $f(x) = x^2$ is $2x$. This derivative can be written as $f'(x) = 2x$, or using the notation $y = x^2$ it can also be written as either $dy/dx = 2x$ or $y' = 2x$.

In differentiating a function $f(x)$, we can use the four-step procedure described in the previous section. As we noted in Section 5.3, however, this procedure is both time-consuming and tedious. More efficiently, Definition 5.1 is used to generate simple general rules for calculating derivatives. These rules are then used to find specific function derivatives as needed (the interested reader can refer to almost any calculus text for a derivation of the rules that follow from this definition). The rules themselves are as follows.

> **Rule 1:** The derivative of a constant is 0. That is, if $f(x) = c$, where c is a real number, $f'(x) = 0$.

This rule is straightforward and, with some thought, obvious. The graph of the equation $y = f(x) = c$ is a straight line parallel to the x-axis. The rate of change of y as x changes, that is the derivative of a line parallel to the x-axis at each point x, is zero. This is because the y value is the same for all values of x, as the y value never changes.

Example 1 What is the derivative of the function $f(x) = 1$?

Solution Using Rule 1 with $c = 10$, we obtain $f'(x) = 0$.

Example 2 What is the derivative of the function $f(x) = -255$?

Solution Using Rule 1, $f'(x) = 0$.

> **Rule 2:** The derivative of the function $f(x) = x^n$, where n is any real number, is
> $$f'(x) = nx^{n-1}.$$

This rule provides the derivative of a class of functions having the form x^n. Examples of such functions are x^2, x^{10}, x^{-8}, $x^{-1/2}$, and $x^{3.1}$. The rule states that if we are given this type of function, the derivative is the power times x raised to one less power.

Example 3 **a.** What is the derivative of $f(x) = x^2$ and

  b. What is the value of the derivative at $x = 3$

Solution **a.** $f(x) = x^2$ is of the form x raised to a power. Using Rule 2 with $n = 2$, we obtain $f'(x) = 2x$.

  b. $f'(3) = 2(3) = 6$

Example 4 **a.** What is the derivative of $f(x) = x^5$?

  b. What is the value of the derivative at $x = 2$

Solution **a.** Using Rule 2 with $n = 5$ we obtain $f'(x) = 5x^4$.

  b. $f'(2) = 5(2)^4 = 5(16) = 80$

Example 5 **a.** Determine the derivative of $f(x) = x^{-2}$.

  b. What is the value of the derivative at $x = 4$?

Solution **a.** The function $f(x) = x^{-2}$ has the form x^n, where $n = -2$. Applying Rule 2, $f'(x) = -2x^{-3}$.

     b. $f'(4) = -2(4)^{-3} = -2\left(\dfrac{1}{4^3}\right) = -2\left(\dfrac{1}{64}\right) = -\dfrac{1}{32}$

Example 6 Determine the derivative of $f(x) = x^{-1/2}$

Solution Using Rule 2, with $n = -$ ½, we obtain $f'(x) = -\dfrac{1}{2}x^{-\frac{3}{2}}$

Example 7 Determine the derivative of $f(x) = x^{3.1}$

Solution Using Rule 2, $f'(x) = 3.1x^{2.1}$.

    If all functions were polynomial of the form x^n, then Rule 2 would be sufficient to handle them. But many functions are not polynomials and cannot be differentiated using either Rules 1 or 2. Thus, if you blindly apply Rule 2, for instance, to any function, you will invariably obtain the wrong derivative.

    One extremely important non-polynomial function that is frequently used in the business, social, and scientific fields is the exponential function. This function was presented in Section 3.5, and the derivative of this function is provided by Rule 3.

Rule 3 If $f(x) = e^{nx}$, where n is any real number, $f'(x) = ne^{nx}$

    The function e^{nx} is indeed a remarkable function. The derivative of the function is a multiple of the function itself; it is impervious to change by differentiation. Geometrically, this signifies the slope of the tangent line at any point on the function is a multiple of its functional (y) value. Another way of looking at this is that if $n > 1$, the function is increasing at a faster and faster rate as x increases.

Example 8 Determine the derivative of the function $f(x) = e^{10x}$.

Solution Using Rule 3 we have $f'(x) = 10e^{10x}$.

Example 9 Determine the derivative of the function $f(x) = e^x$.

Solution Using Rule 3 we have $f'(x) = e^x$. This is a remarkable result. It states that the value of the function e^x and its rate of change at any point x *are identical.*

Example 10 Determine the derivative of the function $f(x) = e^{-5.2x}$.

Solution Using Rule 3 we have $f'(x) = -5.2e^{-5.2x}$.

Example 11 Determine the derivative of the function $f(x) = e^{0.28x}$.

Solution Using Rule 3 we have $f'(x) = 0.28e^{0.28x}$.

Table 5.2 Summarizes the derivatives presented in Rules 1 through 3.

TABLE 5.2 Derivatives of Three Specific Types of Functions.

If f(x) is	It Derivative is	Examples	Comments
c	0	If $f(x) = 10$ $\quad f'(x) = 0$	Rule 1: The derivative of a constant is 0
x^n	nx^{n-1}	If $f(x) = x^6$ $\quad f'(x) = 6x^5$	Rule 2: with $n = 6$ and $n - 1 = 5$
e^{nx}	$n\,e^{nx}$	If $f(x) = e^{3x}$ $\quad f'(x) = 3e^{3x}$	Rule 3: with $n = 3$

Having the derivatives of the functions listed in Table 5.2, we now turn to find the derivatives of combinations of functions. The first such combination is provided by Rule 4.

Rule 4: If $f(x) = cg(x)$, where c is any real number and $g(x)$ has the derivative $g'(x)$, then
$$f'(x) = cg'(x).$$

Note that this is a general rule that applies to many functions. It states that the derivative of a number times a function is simply the number times the derivative of the function.

Example 12 Find $f'(x)$ if $f(x) = 20x^5$.

Solution The function $f(x) = 20x^5$ has the form $cg(x)$, where c is the number 20 and $g(x)$ is the function x^5. Thus, we can use Rule 4 to find its derivative. The derivative of x^5 is $5x^4$. (Why?) Then, using Rule 4, $f'(x) = 20(5x^4) = 100\,x^4$.

Example 13 Find $f'(x)$ if $f(x) = 6x^9$.

Solution Using Rule 4, $f'(x) = 6(9x^8) = 54x^8$.

Example 14 Find $f'(x)$ if $f(x) = 8x^{-2}$.

Solution Using Rule 4, $f'(x) = 8(-2x^{-3}) = -16x^{-3}$.

A question that frequently arises is, Why is $g(x)$ used in Rule 4, rather than the statement, The derivative of cx^n is c times nx^{n-1}? The reason is that Rule 4 is more general and applies to any function whose derivative exits., such as e^{nx}.

Example 15 Find $f'(x)$ if $f(x) = 10e^{5x}$.

Solution Using Rule 4, $f'(x) = 10(5e^{5x}) = 50e^{5x}$.

Exercises 5.4

Determine $f'(x)$ for the functions given in Exercises 1 through 24.

1. $f(x) = 5$ **2.** $f(x) = 10$ **3.** $f(x) = 1/2$ **4.** $f(x) = 2{,}8205$

4. $f(x) = x^2$ **6.** $f(x) = x^3$ **7.** $f(x) = x^4$ **8.** $f(x) = x^5$

9. $f(x) = x^{16}$ **10.** $f(x) = x^{35}$ **11.** $f(x) = x^{118}$ **12.** $f(x) = x^{1805}$

13. $f(x) = x^{-2}$ **14.** $f(x) = x^{-3}$ **15.** $f(x) = x^{-4}$ **16.** p. $f(x) = x^{-7}$

17. $f(x) = x^{-16}$ **18.** r. $f(x) = x^{-21}$ **19.** $f(x) = x^{-236}$ **20.** t. $f(x) = x^{-2378}$

21. $f(x) = x^{1/2}$ **22.** $f(x) = x^{5/2}$ **23.** $f(x) = x^{1/3}$ **24.** $f(x) = x^{7/3}$

Determine the values of $f(2)$ and $f'(2)$ for the functions given in Exercises 25 through 30.

25. $f(x) = x^2$ **26.** $f(x) = x^3$ **27.** $f(x) = 2x^4$

28. $f(x) = 4x^{-2}$ **29.** $f(x) = 7x$ **30.** $f(x) = 10$

Determine $f'(x)$ for the functions given in Exercises 31 through 46.

31. $f(x) = e^x$ **32.** $f(x) = 9e^x$ **33.** $f(x) = 15e^x$ **34.** $f(x) = 21e^x$

35. $f(x) = e^{5x}$ **36.** $f(x) = e^{7x}$ **37.** $f(x) = e^{3x}$ **38.** $f(x) = e^{9x}$

39. $f(x) = e^{-7x}$ **40.** $f(x) = e^{-3x}$ **41.** $f(x) = e^{-2x}$ **42.** $f(x) = e^{-9x}$

43. $f(x) = -2e^{-7x}$ **44.** $f(x) = 4e^{-3x}$ **45.** $f(x) = -5e^{-2x}$ **46.** $f(x) = 3e^{-9x}$

Determine dy/dx for the functions given in Exercises 47 through 66.

47. $y = x^{-1/2}$ **48.** $y = x^{-1/3}$ **49.** $y = x^{-3/8}$ **50.** $y = x^{-7/3}$

51. $y = 2x^8$ **52.** $y = 3x^4$ **53.** $y = 5x^3$ **54.** $y = 4x^5$

55. $y = 3x^6$ **56.** $y = 10x^7$ **57.** $y = 11x^2$ **58.** $y = 36x^2$

59. $y = 7x$ **60.** $y = 10x$ **61.** $y = 5x$ **62.** $y = 9x$

63. $y = 3x^{-2}$ **64.** $y = 6x^{-3}$ **65.** $y = 2x^{-4}$ **66.** $y = 2x^{-7}$

5.5 ADDITION AND SUBTRACTION RULES

If a function can be expressed as either the sum and/or difference of two other functions, Rules 5a and 5b can be applied.

Rule 5a (Addition Rule): If the functions $g_1(x)$ and $g_2(x)$ both have derivatives, then the derivative of the function $f(x) = g_1(x) + g_2(x)$ is:
$$f'(x) = g_1'(x) + g_2'(x).$$

Interpretation: This rule, like Rule 4, is a general rule that applies to many functions. It states that the derivative of a sum of two functions is simply the sum of the derivatives of each function by itself.

Example 1 Determine the derivative of the function $f(x) = 3x^6 + 5$.

Solution The given function is the sum of the two functions $3x^6$ and 5. Here $g_1(x) = 3x^6$, so $g_1'(x) = 18x^5$ (Rule 4), and $g_2(x) = 5$, so $g_2'(x) = 0$ (Rule 1). It now follows from Rule 5 that $f'(x) = 18x^5 + 0 = 18x^5$.

Rule 5a can be extended to include the sum of any number of terms. Furthermore, if we replace the plus signs in Rule 5a with minus signs, we obtain:

Rule 5b (Subtraction Rule): If the functions $g_1(x)$ and $g_2(x)$ both have derivatives, then the derivative of the function $f(x) = g_1(x) - g_2(x)$ is:
$$f'(x) = g_1'(x) - g_2'(x).$$

As with the addition rule (Rule 5a), the subtraction rule also can be extended to include the difference of any number of terms.

Example 2 Find $f'(x)$ for $f(x) = 2x^2 - 3x - 5$.

Solution The derivative of $f(x)$ is the difference of the derivatives of the three terms x^2, $3x$, and 5. Here,

1. $g_1(x) = 2x^2$, hence $g_1'(x) = 4x$ (Rule 4),

2. $g_2(x) = 3x$, hence $g_2'(x) = 3$ (Rule 4),

3. $g_3(x) = 6$, hence $g_3'(x) = 0$ (Rule 1).

Therefore $f'(x) = g_1'(x) - g_2'(x) - g_2'(x) = 4x - 3 - 0 = 4x - 3$.

Example 3 Determine the derivative of the function

$$f(x) = 5x^5 + 6x^3 - 3x + 5 - 8e^x$$

Solution Differentiating $f(x)$ term-by-term, we obtain

$$f'(x) = 25x^4 + 18x^2 - 3 - 8e^x.$$

When the derivative of a function is needed at a particular point, it is obtained by first finding the derivative and then evaluating it at the point of interest. The notation for the derivative of $y = f(x)$ evaluated at $x = x_0$ is either

$$f''(x_0)$$

or

$$\frac{dy}{dx}\bigg|_{x_0}$$

Both symbols are read "the derivative of $y = f(x)$ evaluated at the point $x = x_0$"

Example 4 Find the derivative of $f(x) = 2x^2 - 3x - 6$ at both $x = 5$ and $x = -2$.

Solution First, we determine the derivative of $f(x)$, which is (see Example 2)

$$f'(x) = 4x - 3 - 0 = 4x - 3.$$

Evaluating this derivative at the required points, we obtain

$$f'(5) = 4(5) - 3 = 17$$

and

$$f'(-2) = 4(-2) - 3 = -11$$

Warning: $f'(5)$ denotes the "derivative of $f(x)$ evaluated at $x = 5$" and *not* the "derivative of $f(5)$", which is zero. That is, the value $x = 5$ is substituted into the expression for the derivative $f'(x)$ and not the function $f(x)$. The derivative is taken first and then evaluated; the function is *not* evaluated first and then differentiated.

Table 5.3 Summarizes the rules for finding the derivatives of combinations of functions.

TABLE 5.3 General Rules for Finding the Derivatives of Combinations of Functions.

Combination Type	General Format	Derivative	Rule
A constant times a function	$f(x) = c\, g(x)$	$f(x) = c\, g'(x)$	Rule 4
Addition	$f(x) = g_1(x) + g_2(x)$	$f(x) = g_1'(x) + g_2'(x)$	Rule 5a
Subtraction	$f(x) = g_1(x) - g_2(x)$	$f(x) = g_1'(x) - g_2'(x)$	Rule 5b

Having general rules for finding the derivatives of functions that are added and/or subtracted, in the next section rules for finding the derivatives of functions that are multiplied and/or divided are presented.

Exercises 5.5

Determine $f'(x)$ for the functions given in Exercises 1 through 10.

1. $f(x) = 3x^5 + 4x^2$

2. $f(x) = 4x^3 + 4x^2 + 7x$

3. $f(x) = 6x^3 + 2x^2 + 14x$

4. $f(x) = 5x^5 + 4x^3 + 12$

5. $f(x) = 10x^4 + 5x^2 + 7x$

6. $f(x) = 4x^3 + 4x^2 + 15x + 18$

7. $f(x) = 3x^5 - 4x^2$

8. $f(x) = 15x^2 - 7x + 2$

9. $f(x) = 4x^3 + 2x^2 - 14x$

10. $f(x) = 5x^5 - 4x^3 + 1255$

Determine dy/dx for the functions given in Exercises 11 through 18.

11. $y = 10x^{-4} + 6x^{-2} + 3x$

12. $y = 2x^{-3} + 3x^{-2} + 7x + 5$

13. $y = 3x^{-5} - 4x^{-2}$

14. $y = 2x^{-3} - x^{-2} - 75x$

14. $y = 6x^{-3} + 2x^{-2} - 14x$

16. $f(x) = 10x^4 - 5x^{-2} + 7x^{-1} + 8x + 12$

17. $y = 3x^{-5} - 4x^{-3} + 12$

18. $y = 6x^3 - 12x^{-3} - 9x^{-2} + 7x^{-1} + 15x + 1$

Determine $f'(x)$ for the functions given in Exercises 19 through 32.

19. $f(x) = 3x^5 + 4e^x$

20. $f(x) = 4x^3 + 6e^{5x}$

21. $f(x) = 6e^{3x} + 7e^{2x} + 14x$

22. $f(x) = 5e^x + 4e^{3x} + 12$

23. $f(x) = 5e^{6x} + e^x$

24. $f(x) = 4x^3 + 4e^x + 15e^{3x}$

25. $f(x) = 6x^{-3} + 2e^{-2x}$

26. $f(x) = 5e^{5x} - 4e^{-2}x$

27. $f(x) = 10e^{-4x} + 5e^{3x} + 7x$

28. $f(x) = 4e^{-3x} + 4e^{2x} + 18$

29. $f(x) = 3e^{5x} - 4x^{-2}$

30. $y = 2e^{2x} - 5e^x$

31. $f(x) = 6e^{3x} + 2e^{-2x} - 14x$

32. $f(x) = 10e^{4x} - 5e^{3x} + 3e^{2x} + 8e^x + 12$

33. A company's sales are known to be related to advertising expenditures by the equation $S(x) = 150,000 + 6000x - 50x^2$, where x denotes the monthly advertising expenditures in thousands of dollars.

    a. Find the value of the derivative in sales with respect to advertising expenditures.

    b. Using your answer to part a, determine whether an increase in advertising would increase sales if the present advertising budget is $45,000. Would the situation be different with a $60,000 advertising budget?

34. A firm's total sales, in millions of dollars, is given by the equation $R(x) = 3x + 1/2\ x^2$, where x denotes the number of years the firm has been in operation.

    a. Determine the firm's average rate of growth in sales for its first 7 years in business.

    b. Determine the firm's instantaneous rate of growth after its seventh year of business.

    c. What will the firm's total sales be at the end of the tenth year if sales continue to grow *at the same rate of growth* achieved after the seventh year?

    d. What total sales has the firm actually reached after its tenth year of business? Compare your answer to the answer for part c. and explain the discrepancy.

35. Figure 5.23 illustrates the distance traveled as a function of time for Ms. Williams' 2-mile trip through the Holland Tunnel. The equation of the curve illustrated is $D = -0.05t^3 + 0.25t^2 + 0.3t$, where D is the distance traveled, in mile, and t is measured in minutes from the start of the trip.

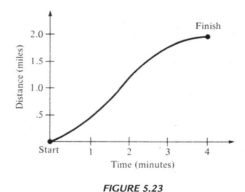

FIGURE 5.23

    a. Determine the average rate of change in distance with respect to time for the complete 2-mile trip. Physically, what does your answer represent?

    b. Redraw Figure 5.23 on a separate sheet of graph paper. Draw a line on the graph whose slope represents the average speed determined in part a.

c. Determine the speed of Ms. Williams' car at exactly 1 minute after the start his trip. (*Hint:* Speed is the instantaneous rate of change in distance traveled with respect to time.)

d. Draw a line on the graph constructed for part b, whose slope represents the instantaneous speed determined in part c.

e. Determine the exact speed that would be indicated on Ms. Williams' speedometer as he emerges from the tunnel ($t = 4$).

5.6 PRODUCT AND QUOTIENT RULES

In addition to functions being added and subtracted, as presented in the previous section, functions can also be multiplied, divided, and raised to a power. This section shows how to find the derivatives of functions that are multiplied or divided.

Rule 6 (Product Rule): If the functions $g_1(x)$ and $g_2(x)$ both have derivatives, then the derivative of the function $f(x) = g_1(x)\,g_2(x)$ is:

$$f'(x) = g_1(x)\,g'_2(x) + g_2(x)\,g'_1(x)$$

This rule states how to find the derivative of two functions that are multiplied together. It says *that the derivative of a product of two functions is the first function times the derivative of the second function plus the second function times the derivative of the first function* (note that the derivative of a product is *not* equal to the product of the individual derivatives).

Example 1 Find $f'(x)$ if $f(x) = x^2 e^x$.

Solution We first note that the function whose derivative we want is the product of two functions, which in this case are x^2 and e^x. Thus, $g_1(x) = x^2$ and $g_2(x) = e^x$, and their associated derivatives are $g'_1(x) = 2x$ (from Rule 2) and $g'_2(x) = e^x$, (from Rule 3), respectively. Using Rule 6, we obtain

$$f'(x) = x^2 e^x + e^x(2x) = x^2 e^x + 2x e^x.$$

Example 2 Determine the derivative of the function $f(x) = x^2(x^3 + 3)$.

Solution We can do this problem in two ways.

First Way: Use the Product Rule. Because $f(x)$ is the product of two functions, x^2 and $(x^3 + 3)$, we set $g_1(x) = x^2$ and $g_2(x) = (x^3 + 3)$. The respective derivatives of these functions are $g'_1(x) = 2x$ and $g'_2(x) = 3x^2$, respectively. Using Rule 6 we find

$$f'(x) = x^2(3x^2) + (x^3 + 3)(2x)$$

$$= 3x^4 + 2x^4 + 6x$$

$$= 5x^4 + 6x.$$

Second Way: The same answer is obtained by algebraically simplifying the original function to $f(x) = x^2(x^3 + 3) = x^5 + 3x^2$, and then differentiating term-by-term using Rule 5a.

At this point, we have rules for determining the derivative of functions that are added, subtracted, and multiplied. The next rule provides the details of determining the derivative of two functions that are divided.

Rule 7 (Quotient Rule): If the functions $g_1(x)$ and $g_2(x)$ both have derivatives, with $g_2(x) \neq 0$, then the derivative of the function $f(x) = g_1(x)/g_2(x)$ is:

$$f'(x) = \frac{g_2(x)g'_1(x) - g_1(x)g'_2(x)}{\left[g_2(x)\right]^2}$$

This rule states how to determine the derivative of a function consisting of one function divided by another. It says *the derivative of a quotient is the function in the denominator times the derivative of the function in the numerator minus the numerator function times the derivative of the denominator function, with the entire result divided by the square of the denominator function* (note that the derivative of a quotient is *not* equal to the quotient of the individual derivatives).

Example 3 Find the derivative of $f(x) = e^x / x^5$.

Solution We first note that the derivative we want is the quotient of two functions, which in this case are e^x and x^5. Thus, $g_1(x) = e^x$ and $g_2(x) = x^5$, and their associated derivatives are $g_1'(x) = e^x$ (from Rule 3) and $g'_2(x) = 5x^4$ (from Rule 2), respectively. Substituting the appropriate terms into Equation 5.10, we obtain

$$f'(x) = \frac{x^5(e^x) - e^x(5x^4)}{\left[x^5\right]^2} = \frac{x^5 e^x - 5x^4 e^x}{x^{10}} = \frac{xe^x - 5e^x}{x^6}, \quad x \neq 0$$

Example 4 Find $f'(x)$ for $f(x) = e^x/(x^2 + 4)$.

Solution Again, we first notice that it is a quotient whose derivative is desired, which means we will use the quotient rule (Rule 7). Here, the numerator function is $g_1(x) = e^x$ and the denominator function is $g_2(x) = (x^2 + 4)$. Thus, $g'_1(x) = e^x$ (from Rule 3) and $g'_2(x) = 2x$ (from Rule 2). Using Rule 7, we obtain

$$f'(x) = \frac{(x^2 + 4)(e^x) - e^x(2x)}{\left[x^2 + 4\right]^2},$$

which, after simplifying, becomes

$$f'(x) = \frac{(x^2 - 2x + 4)e^x}{\left[x^2 + 4\right]^2},$$

Exercise 5.6

Determine $f'(x)$ for the functions given in Exercises 1 through 6.

1. $f(x) = (8x + 2)(6x^3 - 12x^{-3})$

2. $f(x) = 6x^3(12x^5 - 9x^2)$

3. $f(x) = (x^5 + 3)e^x$

4. $f(x) = (3x^{-5} - 4x)(2x^3 + 6x)$

5. $f(x) = 9e^x(x^2)$

6. $f(x) = (3x^{-5} - 4x^{-2})(5x^5 + 4x^3 + 12)$

Determine $f'(x)$ for the functions given in Exercises 7 through 12.

7. $f(x) = e^x(x^5)$

8. $f(x) = (x^3 + 6x^2)e^x$

9. $f(x) = (x^4)(6x^{-3} + 2e^{-2x})$

10. $f(x) = (6x^{-3} + 2x^{-2} - 14x)(3x^5 - 4x + 109)$

11. $f(x) = 3x^7(e^{7x})$

12. $f(x) = (8x + 12)(6x^3 - 12x^{-3})$

Determine $f'(x)$ for the functions given in Exercises 13 through 19.

13. $f(x) = e^x/(x^5 + 3)$

14. $f(x) = (2x^3 + 6x)/(3x^{-5} - 7)$

15. $f(x) = (x^4)/(6x^{-3} + 2e^{-2x})$

16. $f(x) = (6x^{-3} + 2x^{-2} - 14x)/(3x^5 - 4x + 109)$

17. $f(x) = 3x/(e^{4x})$

18. $f(x) = (8x + 12)/(2x^3 - 12)$

19. $f(x) = (3x^{-4} + 2e^{-2x})/x^4$

Determine $f'(x)$ for the functions given in Exercises 20 through 26.

20. $f(x) = (2x^{-2} - 14x)/(3x^6 - 25)$

21. $f(x) = 5e^{3x}/14x^3$

22. $f(x) = (8x^2 + 10x)/(6x^5 - 12x^{-1})$

23. $f(x) = (10x^{-2} - 15x)/5$

24. $f(x) = x^5/5 - x^3/3 - x^2/2 + 10$

25. $f(x) = (x^{10} + e^x)/2$

26. $f(x) = 1/x^5 + 1/x^2$

5.7   ADDITIONAL RULES

Occasionally, we have to differentiate a function raised to a power such as $f(x) = (x^2 + 3x + 12)^3$. The following rule provides the method for finding the derivative of this type of function.

Rule 8 (Power Rule): If $f(x) = [g(x)]^n$ where n is a fixed real number and the derivative of $g(x)$ exists, then
$$f'(x) = n[g(x)]^{n-1} g'(x)$$

Interpretation: Although this rule resembles Rule 2, it is quite different. Rule 2 deals with a single variable raised to a power, as in x^n, where this rule deals with a function that may be made up of many terms raised to a power, such as $(x^2 + 3x + e^x)^3$. The rule states that to find the derivative of a function raised to a power, first treat the function inside the brackets, denoted here by $g(x)$, as a single term and apply Rule 2. Then multiply this result by $g'(x)$, which is simply the derivative of what is inside the brackets.

Example 1 Determine the derivative of $f(x) = (x^2 + 3x + 12)^3$.

Solution First notice that the function consists of the quantity $(x^2 + 3x + 12)$ raised to the third power. As such, it is a function raised to a power, whose derivative can be found using the Power Rule (Rule 8). Therefore, $g(x) = x^2 + 3x + 12$, which is the quantity being raised to a power, and the power itself is 3. Thus, $n = 3$ and $g'(x) = 2x + 3$, and it follows from Rule 8 that $f'(x) = 3(x^2 + 3x + 12)^2(2x + 3)$.

This answer can also be obtained by a second procedure, which firsts cube the quantity $(x^2 + 3x + 12)$, and then takes the derivative of each term using the Addition Rule (Rule 5a). Using Rule 8, however, is quicker.

Example 2 Differentiate $f(x) = (3 + 4e^{7x})^{1.95}$

Solution First, notice that $f(x)$ is the function $g(x) = (3 + 4e^{7x})$, raised to the 1.95 power. Using Rule 8 in combination with other appropriate rules, we find $g'(x) = 0 + 4(7)e^{7x} = 28e^{7x}$. It then follows from the power rule that

$$f'(x) = 1.95\,(3 + 4e^{7x})^{1.95-1}\,(28e^{7x}) = 54.6\,(3 + 4e^{7x})^{.95}\,(e^{7x})$$

Example 3 Differentiate $f(x) = \sqrt{x^2 + 1}$

Solution Because $\sqrt{x^2 + 1} = (x^2 + 1)^{1/2}$ we can use Rule 8. Here we set $g(x) = (x^2 + 1)$ and take $n = \frac{1}{2}$. Then $g'(x) = 2x$, and $f'(x) = \frac{1}{2}\,(x^2 + 1)^{(1/2 - 1)}\,(2x) = x(x^2 + 1)^{-1/2}$.

Note that in this case, the second procedure used in Example 1 is not applicable for this function, and Rule 8 provides the only procedure for finding the derivative.

To this point, we have considered only functions given by the one equation $y = f(x)$. Frequently, however, business problems are modeled by a set of two equations having the form $y = g_1(u)$ and $u = g_2(x)$. An example of such a set is $y = u^2 + 3u$ and $u = 6x + 2$. Because we can substitute the value for u into the equation for y, it follows that y is ultimately a function of x, and it makes sense to ask for the derivative dy/dx. Interestingly, we can find dy/dx directly without substituting first, if the individual derivatives of the functions $g_1(u)$ and $g_2(x)$ are known. The procedure is known as the chain rule.

Rule 9 (Chain Rule): If $y = g_1(u)$ and $u = g_2(x)$, and if both the derivatives dy/du and du/dx are known, then

$$\frac{dy}{dx} = \left(\frac{dy}{du}\right)\left(\frac{du}{dx}\right)$$

This rule states that the desired derivative of y as a function of x, dy/dx, can be found by multiplying the first derivative, dy/du by the second derivative, du/dx. The term *chain rule* evolves from the ultimate derivative, dy/dx, being a "chain" of two other derivatives, namely, dy/du and du/dx.

Example 4: Determine dy/dx if $y = u^2 - 5u + 7$ and $u = 7x^2 + 10$.

Solution Using the chain rule, we first determine

$$\frac{dy}{du} = 2u - 5 \qquad and \qquad \frac{du}{dx} = 14x$$

Then,

$$\frac{dy}{dx} = (2u - 5)(14x) = \left[2(7x^2 + 10) - 5\right](14x) = 196x^3 + 210x$$

Alternatively, we could first substitute the expression for u into the equation for y, obtaining

$$y = (7x^2 + 10)^2 - 5(7x^2 + 10) + 7 = 49x^4 + 105x^2 + 57.$$

Then, differentiating this expression directly, we again determine

$$\frac{dy}{dx} = 196x^3 + 210x$$

Example 5 A firm's monthly sales revenue S is known to be related to its advertising expenditures by the equation $S = 100 - 30E + 0.2E^2$, where E represents monthly advertising expenditures in thousands of dollars. The company allocates its monthly advertising expenditure based on the equation $E = 0.1x^2 - 0.3x + 5000$, where x represents thousands of units sold the previous month. Determine the derivative dS/dx, which is the instantaneous rate of change of sales with respect to previous monthly sales.

Solution We are asked to find dS/dx, given $S = 100 - 30E + 0.2E^2$ and $E = 0.1x^2 - 0.3x + 5000$. Modifying the notation in Rule 9 so that it is applicable to this situation, we obtain

$$\frac{dS}{dx} = \frac{dS}{dE}\frac{dE}{dx}$$

Performing the appropriate operations, we have

$$\frac{dS}{dx} = (-30 + 0.4E)(0.2x - 0.3)$$

which, upon substituting $0.1x^2 - 0.3x + 5000$ for E, becomes

$$\frac{dS}{dx} = 0.008x^3 - 0.036x^2 + 394.036x - 591$$

The alternative procedure suggested in Example 4 is applicable here also.

Table 5.4 summarizes all of the general rules for finding the derivatives of combinations of functions presented in this and the previous sections.

TABLE 5.4 General Rules for Finding Derivatives of Combinations of Functions.

Combination Type	General Format	Derivative
A constant times a function	$f(x) = c\, g(x)$	$f'(x) = c\, g'(x)$
Addition	$f(x) = g_1(x) + g_2(x)$	$f'(x) = g_1'(x) + g_2'(x)$
Subtraction	$f(x) = g_1(x) - g_2(x)$	$f'(x) = g_1'(x) - g_2'(x)$
Multiplication	$f(x) = g_1(x) \cdot g_2(x)$	$f'(x) = g_1'(x)\, g_2'(x) + g_2(x)\, g_1'(x)$
Division	$f(x) = g_1(x) / g_2(x)$	$f'(x) = [g_2(x)\, g_1'(x) - g_1(x)\, g_2'(x)]/[g_2(x)]^2$
Power	$f(x) = [g(x)]^n$	$f'(x) = n[g(x)]^{n-1}\, g'(x)$
Chained	$y = g_1(u)$ and $u = g_2(x)$	$dy/dx = (dy/du)(du/dx)$

Exercises 5.7

Determine $f'(x)$ for the functions given in Exercises 1 through 4.

1. $f(x) = (6x^{-3} + 2e^{-2x})^3$

2. $f(x) = (5e^{5x} - 4e^{2x})^5$

3. $f(x) = (6e^{-4x} + 7x)^2$

4. $f(x) = (4e^{2x} + 9x)^4$

Determine $f'(x)$ for the functions given in Exercises 5 through 8.

5. $f(x) = (2e^{9x} - 6x^{-2})^8$

6. $f(x) = (e^{2x} - 15e^x)^3$

7. $f(x) = (10x^3 + 2e^{-2x})^{15}$

8. $f(x) = (2x^{-3} - x^4 + 12)^7$

Using the chain rule, determine dS/dx for the functions given in Exercises 9 through 11.

9. $S = E^2 + 3E$, where $E = x^2$

10. $S = E^2 + 3$, where $E = x^3 + 5x^2$

11. $S = E/(E + 2)$, where $E = 3x^2 + 5$

5.8 HIGHER-ORDER DERIVATIVES

Just as we took the first derivative of a differentiable function, we usually can differentiate the derivative itself. The derivative of a first derivative is called the *second derivative* and is commonly denoted by $f''(x)$, y', d^2y/dx^2, or $d^2[f(x)]/dx^2$. To obtain the second derivative of a function, we treat the first derivative as a new function and differentiate it using the rules provided in the previous sections. Generally, only the first and second derivatives of a function have applications to business problems.

Example 1 Determine the first derivative and second derivative of the function $y = 3x^2 + 2x + 5$.

Solution We initially determine the first derivative and then take the derivative of this function to obtains the second derivative. The first derivative is obtained, using Rule 5a, as

$$f'(x) = \frac{dy}{dx} = 6x + 2$$

The derivative of this function, which is the second derivative, is obtained using Rule 5a once again, as

$$f''(x) = \frac{d^2y}{dx^2} = 6$$

Example 2 Determine d^2y/dx^2 for the function $y = x^2e^x$.

Solution We first find dy/dx and then differentiate it to obtain d^2y/dx^2. Using the product rule (Rule 6), we have

$$dy/dx = x^2e^x + 2xe^x$$

Differentiating dy/dx term-by-term, which requires the using product rule (Rule 6) for each of the two terms in the sum, we obtain:

$$\frac{d^2y}{dx^2} = (x^2e^x + 2xe^x) + (2xe^x + 2e^x)$$

$$= x^2e^x + 4xe^x + 2e^x$$

$$= e^x(x^2 + 4x + 2)$$

Continuing in this manner, we could define third-, fourth-, fifth-, and higher-order derivatives. However, only the first two derivatives of a function have applications to business problems. We consider some of these applications in Chapter 6.

Exercises 5.8

In Exercises 1 through 6, find dy/dx and d^2y/dx^2 and evaluate the second derivative at the points $x = 1$ and $x = 2$.

1. $y = x^5 + 4x^2 + 3x$.   2. $y = x + 3$.

3. $y = x^4 + 3x^2 + 4x +$ 4. $y = e^x$.

5. $y = x^3e^{10x}$.   6. $y = 5x^2e^{7x}$.

7. Figure 5.24 represents the distance traveled as a function of time for Mr. Williams' 2-mile trip through the Holland Tunnel, first considered in Exercise 27 in Section 5.5. The equation of the curve illustrated is $D = -0.05t^3 + 0.25t^2 + 0.3t$, where D is the distance traveled in miles, and t is measured in minutes from the start of the trip.

   a. Determine a general expression for the speed of Mr. Williams' car at any time during the trip.

   b. Determine the speed that would be indicated on Mr. Williams' speedometer at $t = 1$ minute.

   c. Determine a general expression for Mr. Williams' acceleration at any time during the trip. (*Hint:* Acceleration $= d(speed)/dt = d^2D/dt^2$.)

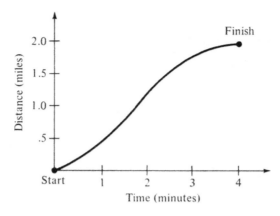

FIGURE 5.24

d. Using your answer for part c., determine Mr. Williams' acceleration at $t = 1$ minute and $t = 3$ minutes. What is the significance of the negative sign at $t = 3$ minutes?

5.9 SUMMARY OF KEY POINTS

Key Terms

- Average rate of change
- Chain rule
- Continuous
- Corner
- Derivative
- Differentiable function
- Discontinuous
- Instantaneous rate of change
- Limit
- Limit from the left
- Limit from the right
- Marginal cost
- Marginal revenue
- Secant line
- Tangent line

Key Concepts

Rates of Change

- The average rate of change of the function $y = f(x)$ over the interval $[x_1, x_2]$ is the slope of the line that passes through the two point (x_1, y_1) and (x_2, y_2), where $y_1 = f(x_1)$ and $y_2 = f(x_2)$.
- The instantaneous rate of change of the function $y = f(x)$ at x_1 is the slope of the tangent line to the graph of $f(x)$ at the point (x_1, y_1), where $y_1 = f(x_1)$.

Limits and Continuity

- If a function is continuous at x_1, then its graph must show a seamless transition of functional value as x moves near x_1 from both sides of x_1 and then onto x_1 itself.

The Derivative

- The derivative is the instantaneous rate of change of a function $y = f(x)$.
- To find the derivative of a function at a particular value $x = x_1$, first obtain the derivate of the function $y = f(x)$ and then evaluate the derivative with $x = x_1$.

Differentiable Functions

- A necessary condition for a function to have a derivative at a point is that the function must be continuous at that point. However, this does not ensure that a derivative exists there.
- A function has a derivative at a point if the function is both continuous and does not have a corner at that point.

OPTIMIZATION

In this Chapter

Optimization, which is the process of either maximizing or minimizing a quantity, is central to most business problems. For example, how many units of a product should be manufactured to maximize profit? How often should material be ordered to minimize total inventory cost? What is the best method for government agencies to control money supplies so as to maximize employment?

In Section 6.1, we develop the theory of optimization based on the derivative. Recall from Chapter 5 that differentiation is a mathematical operation performed on functions. To apply the derivative to business situations, we must first construct mathematical equations that realistically model or represent the situations of interest. The importance of mathematical models is presented in Section 6.2

Once the appropriate equations have been obtained that realistically model the situations being investigated, optimization techniques can be applied. Examples of this approach are presented in Sections 6.3 and 6.4.

It is important to realize that not all applications of the derivative deal with optimization. Rates of change have direct applications themselves, many of which were presented in the examples and exercises provided in Chapter 5. An additional, and more advanced example, is presented in Section 6.5.

6.1   OPTIMIZATION THROUGH DIFFERENTIATION

In this section, we introduce a method for finding the high and low values of mathematical functions. The high values are known as *maximum points*, or *maxima*, for short, and the low points as *minimum points*, or *minima*, for short. Together, a function's maxima and minima are referred to as its *optimal values*. In subsequent sections, we show how to derive a number of important commercial functions, and then how to locate these functions' optimal values using the techniques presented in this section.

As a specific example, assume we know that the relationship between profit P (in dollars) and the number of items manufactured and sold, x, for a particular industry is modeled by the equation

$$P = -x^2 + 120x - 600$$

(Eq. 6.1)

We want to know the value(s) of x that will maximize P. That is, how many items should be manufactured if the objective is to maximum profit?

One approach, known as trial and error is to substitute different values of x into Equation 6.1, calculate corresponding values of P, and determine the values of x that maximizes P. For example, substituting $x = 10$ into Equation 6.1, we obtain $P = -(10)2 + 120(10) - 600 = \$500$. Substituting $x = 20$ into Equation 6.1 yields $P = \$1,400$. Continuing in this manner, we generate Table 6.1. It appears that $x = 60$ will produce a maximum profit of $P = \$3000$. But can we be sure? Perhaps $x = 61$ will generate a bigger profit. Or, perhaps the maximum profit occurs at $x = 137$.

TABLE 6.1 Evaluating $P = -x^2 + 120x - 600$ at Various Values of x.

x (units)	0	10	20	40	60	70	80
P ($)	−600	500	1400	2600	3000	2900	2600

The difficulty in calculating a function for only some values of the independent variable, as is done in Table 6.1, is that we are never sure that the optimum does not occur at some other value.

A second and more useful procedure for finding maxima and minima is through graphing. Agreeing that the *maximum value* of a function is the largest value the function can obtain, and similarly that the *minimum value* of a function is the smallest value the function can obtain, we conclude that maxima appear as high points and minima appear as low points on a function's graph.

Plotting the points given in Table 6.1 and realizing that $P = -x^2 + 120x - 600$ is a quadratic equation (see Section 3.4), we draw Figure 6.1 as the graph of Equation 6.1. It is evident from the curve that the maximum profit is $3,000 from a production run of $x = 60$ units. At this time, Equation 6.1 does not have a minimum, since there is no point in Figure 6.1 that is smaller than every other point.

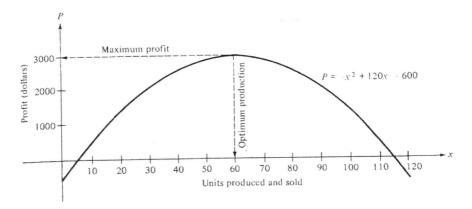

FIGURE 6.1 Graphically determining a maximum point.

Example 1 Graph $y = x^2 - 10x + 16$ and determine the maximum and minimum values for y.

Solution This equation is quadratic. Arbitrarily selecting values of x, computing the corresponding values of y, and plotting these points, we obtain Figure 6.2. The minimum value of y is -9, which occurs when $x = 5$. The function does not have a maximum because there is no point on the graph that is larger than every other point.

In both Figures 6.1 and 6.2, one of the optimal values, either the maximum or the minimum, did not exist. The reason was that the domain, the set of allowable values for the independent variable x, was infinite. The larger we allowed x to become in Figure 6.1, the smaller P became. The larger we allowed x to become in Figure 6.2, the larger y became. Both situations are unrealistic from a business point of view. In business, there is always a limit beyond which it is either impossible or not feasible to go.

On an absolute scale, every product is limited in number by the amount of raw materials available in the world. Realistically, production runs are also limited by the demand for the product, the time required to make the product, and the capital investment necessary to produce the product. Even under the best conditions, only a finite number (perhaps a large finite number) of each product can be produced. Similarly, every service, on an absolute scale, is limited by the time people have available to perform the service.

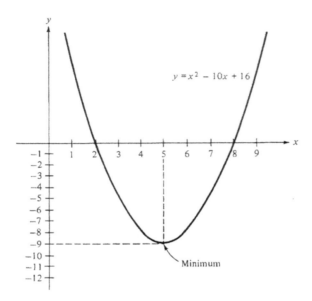

FIGURE 6.2 Graphically determining a minimum point.

Thus, every business process has finite limits or boundaries. Therefore every mathematical model of a business situation must reflect these limits. Most often, this is done by restricting the domain. Equation 6.1 is not a good model for a commercial situation, because it does not indicate limits on the values of x. A more realistic model is

$$P = -x^2 + 120x - 600 \qquad (0 \le x \le 150) \qquad \text{(Eq. 6.2)}$$

We now have the additional information that no less than 0 products and no more than 150 products can be produced. With these restrictions, the graph of Equation 6.2 becomes Figure 6.3, where we have plotted only points associated with the domain $0 \le x \le 150$. It now follows that the maximum is

$P = \$3000$, occurring at $x = 60$, *and* the minimum is $P = -\$5,100$, occurring at $x = 150$.

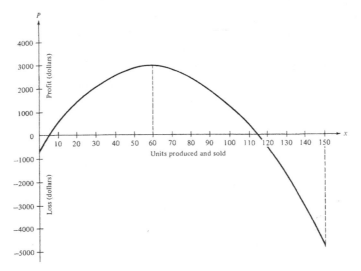

FIGURE 6.3 Endpoints must be considered.

Example 2 Figure 6.4 represents the relationship between the dollar value of inventory on hand for a distribution center over a 12-month time period t, where t is in months ($0 \leq t \leq 12$). Find the maximum and minimum dollar values of inventory on hand.

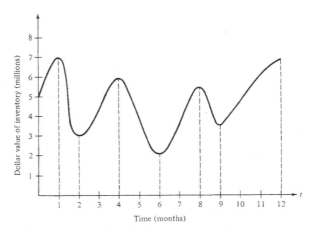

FIGURE 6.4 Multiple locally optimum points can exist.

Solution The maximum value of inventory on hand is $7 million, which occurs at both the first ($t = 1$) and last months ($t = 12$). The minimum value is $2 million, which occurs in the sixth month ($t = 6$).

It follows from Example 2 that optimal values can occur at more than one place. Note that the maximum occurred at both $t = 1$ and $t = 12$.

The peaks at $t = 4$ and $t = 8$ and the valleys at $t = 2$ and $t = 9$ also are interesting. They are not maxima or minima because there are other points on the graph that are higher ($t = 1$) and lower ($t = 6$). Nonetheless, they are high and low points for some small subset of the domain and, as such, are often called *relative maxima* and *relative minima*. Detailed development of their properties is given in the exercises.

Finding optimal points by graphing has two disadvantages. The first is a lack of accuracy. Can we really conclude from Figure 6.2 that the maximum occurs at $t = 6$ and not at $t = 6.1$ or 5.98? No, but as one rarely needs more accuracy than can be obtained from a graph, this disadvantage is not serious. The major difficulty is that the curve must be graphed. Accurately graphing a complicated equation is frequently a difficult and time–consuming procedure that we would like to avoid if possible.

The third approach to optimization, which is more accurate, direct, and eliminates the need for graphing, uses differentiation. This differentiation method is based on the following theorem, the proof of which can be found in most calculus texts.

Theorem 6.1 Let a function be defined on a domain $a \leq x \leq b$ and have a maximum or minimum at a point c, where $a < c < b$. If the derivative of the function exists at $x = c$, then the derivative must be zero there.

This theorem, after a little thought, is rather obvious. Note in Figures 6.2 and 6.3 that the tangents to each curve at the maximum and minimum points, if drawn, would be horizontal, and therefore have zero slopes. Because the derivative is the slope of the tangent line, it follows that the derivative at such maximum or minimum points is zero.

This theorem simplifies the search for optimal points. Maxima and minima can occur only where the first derivative is zero or where the theorem is not applicable. The theorem is not applicable in two cases. First, it says nothing about optimal points at the boundaries of the domain, usually called the *endpoints*. In Theorem 6.1, the endpoints are $x = a$ and $x = b$. Second, the theorem says nothing about optimal points at places where the derivative does not exist. Accordingly, maxima and minima of functions can occur at only one of three points and nowhere else:

1. Points where the first derivative is zero

2. Points where the first derivatives do not exist

3. Endpoints.

To locate maximum and minimum points for any function, first, find all points where the first derivative is zero, then all points where the derivative does not exist, and finally all endpoints. These points, taken together, are referred to as *candidate points*, or *candidates*, for short. Substituting each of the candidate points into the given equation will determine which of the points optimize it.

Example 3 Determine the maximum and minimum values of the function

$$P = -x^2 + 120x - 600 \quad (0 \le x \le 150)$$

Solution Differentiating the given function, we have $dP/dx = -2x + 120$. To find values of x for which this derivative equals zero, we set the derivative equal to zero and solve for x. Accordingly,

$$-2x + 120 = 0$$
$$-2x = -120$$
$$x = 60$$

The first derivative exists everywhere, because for each value of x the quantity $dy/dx = -2x + 120$ is defined; thus, there are no points where the first derivative does not exit (category 2). The end points are $x = 0$, $x = 150$. Thus, the candidates for the points at which the function's maximum and minimum values occur are $x = 60$, $x = 0$, and $x = 150$.

Gathering these candidate points together, we substitute each into the given function and evaluate the corresponding values for P. This is done in Table 6.2. It is now clear that the maximum is $P = 3,000$, which occurs at $x = 60$, and the minimum is $P = -5,100$, which occurs at $x = 150$.

TABLE 6.2 Maximum and Minimum Candidate Points.

x	$P = -x + 120x - 600$
60	3000 <--Maximum
0	−600
150	−5100 <--Minimum

Example 4 Determine the maximum and minimum values of the function

$$y = x^3 - 21x^2 + 120x - 100 \qquad (1 \leq x \leq 12).$$

Solution Differentiating the given function, we find $dy/dx = 3x^2 - 42x + 120$. To locate all values of x for which this derivative is zero, we set the derivative equal to zero and solve for x. Thus,

$$3x^2 - 42x + 120 = 0$$
$$x^2 - 14x + 40 = 0 \qquad \text{(dividing by 3)}$$
$$(x - 10)(x - 4) = 0 \qquad \text{(factoring)}$$
$$x = 10 \text{ and } x = 4.$$

Thus, the first derivative is zero at $x = 10$ and $x = 4$.

There are no points for which the first derivative does not exist, because $dy/dx = 3x^2 - 42x + 120$ can be evaluated at all values of x. The end points for this problem are $x = 1$ and $x = 12$.

Evaluating the given function at the candidate points, $x = 10$, $x = 4$, $x = 1$, and $x = 12$, we obtain Table 6.3. It follows that the maximum is $y = 108$ which occurs at $x = 4$, and the minimum is $y = 0$ which occurs at two places, $x = 10$ and $x = 1$.

TABLE 6.3 Maximum and Minimum Candidate Points.

x	$y = -x^3 + 120x - 100$
1	0 <--Minimum
4	108 <--Maximum
10	0 <--Minimum
12	44

Example 5 Determine the maximum and minimum of $y = t^3 - 9t^2 - 120t + 1{,}500$ $(0 \leq t \leq 30)$.

Solution Differentiating the given function, we find $dy/dt = 3t^2 - 18t - 120$. To locate all values of t for which this derivative is zero, we set the derivative equal to zero and solve for t. Thus

$$3t^2 - 18t - 120 = 0$$
$$t^2 - 6t - 40 = 0 \qquad \text{(dividing by 3)}$$
$$(t + 4)(t - 10) = 0 \qquad \text{(factoring)}$$
$$t = -4 \text{ and } t = 10.$$

The first derivative is zero at $t = -4$ and $t = 10$. Because $t = -4$ is outside our domain and not an allowable point for this problem, we disregard it.

The first derivative exists everywhere because for each value of t the quantity $3t^2 - 18t - 120$ is defined. The endpoints for this problem are at $t = 0$ and $t = 30$. Evaluating the given function at the candidate points, $t = 0$, $t = 10$, and $t = 30$, we obtain Table 6.4. It follows that the maximum is $y = 16{,}800$ occurring at $t = 30$, and the minimum is $y = 400$, which occurs at $t = 10$.

TABLE 6.4 Maximum and Minimum Candidate Points.

x	$y = t^3 - 9t + 1500$
0	1,500
10	400 <--Minimum
30	16,800 <--Maximum

Example 6 Determine the maximum and minimum values of the function

$$D = (P - 20)^2 (P - 30)^2 \qquad (20 \le P \le 35)$$

Solution Differentiating this function with the product rule for differentiation, we obtain

$$\frac{dD}{dP} = (P - 20)^2 \, 2(P - 30) + (P - 30)^2 \, 2(P - 20)$$

$$= 2(P - 20)^2 (P - 30) + 2(P - 30)^2 (P - 20)$$

Setting this quantity equal to zero and solving for P, we have

$$2(P - 20)^2 (P - 30) + 2(P - 30)^2 (P - 20) = 0$$
$$2(P - 20)(P - 30)[(P - 20) + (P - 30)] = 0$$
$$2(P - 20)(P - 30)(2P - 50) = 0.$$
$$P = 20, \quad P = 30, \quad \text{and} \quad P = 25.$$

The derivative exists everywhere, and the endpoints are $P = 20$ and $P = 35$. It follows from Table 6.5, where all of the candidates are evaluated, that the maximum is $D = 5{,}625$, which occurs at $P = 35$, and the minimum is $P = 0$, which occurs at both $P = 20$ and $P = 30$.

TABLE 6.5 Maximum and Minimum Candidate Points.

P	$D = (P - 20)^2 (P - 30)^2$
20	0 <--Minimum
25	625
30	0 <--Minimum
35	5,625 <--Maximum

Exercises 6.1

In Exercises 1 through 5 find the maximum and minimum values of the given functions two ways: First, graph the functions, visually locating the optimal points and, second, use differentiation.

1. $y = 3x^2 - 24x - 5,$ $(0 \leq x \leq 6)$

2. $y = -x^2 + 25x - 3.5,$ $(4 \leq x \leq 16)$

3. $y = 7x^2 + 56x + 25,$ $(0 \leq x \leq 5)$

4. $y = -5x^2 + 100x,$ $(0 \leq x \leq 20)$

5. $y = 2x - 1,$ $(25 \leq x \leq 50)$

In Exercises 6 through 15 use the first derivative to determine maximum and minimum values of the given functions.

6. $D = t^3 - 12t + 7,$ $(-3 \leq t \leq 3)$

7. $D = t^3 - 12t + 7,$ $(-4 \leq t \leq 4)$

8. $D = t^3 - 12t + 7,$ $(-5 \leq t \leq 5)$

9. $D = t^3 - 12t + 7,$ $(0 \leq t \leq 3)$

10. $P = 2n^3 - 54n + 2800,$ $(-3 \leq n \leq 5)$

11. $y = t^3 + 3t^2 - 105t + 20$ $(-10 \leq t \leq 10)$

12. $y = x^3 - 48x + 57,$ $(0 \leq x \leq 4)$

13. $x = (1/3)t^3 - 4t^2 + 12,$ $(0 \leq t \leq 9)$

14. $T = c^3 - 10c^2 + 12c + 10,$ $(0 \leq c \leq 10)$

15. $y = (x - 10)(x - 20),$ $(5 \leq x \leq 25)$

16. A television manufacturer has found that the profit P (in dollars) obtained from selling x television sets per week is given by the formula $P = -x^2 + 300x - 5,000$. Determine how many television sets the manufacturer should produce to maximize profits. What is the maximum profit that can be realized?

17. A manufacturing company has found that the profit P (in dollars) realized in selling x items is given by $P = 22x - \dfrac{x^2}{2,000} - 10,000$

    a. How many items should the manufacturer produce to maximize profit?

    b. Find the maximum profit.

18. The manufacturer described in Exercise 16 also has determined that the total cost of producing x items, denoted by TC (in dollars), is given by

$$TC = \frac{x^2}{2,000} - 7x + 10,000.$$

    a. Determine the number of items that should be produced if the manufacturer's goal is to minimize total cost rather than to maximize profit.

    b. What is the minimum total cost? Is this answer reasonable?

19. In the course of a week, a refrigerator manufacturer can sell x refrigerators at a unit price of $300. The total cost TC (in dollars) is given as $TC = 2.5x^2 - 200x + 20,000$.

    a. How many units should the firm produce each to maximize profits?

    b. Determine the maximum profit.

    c. Determine the total cost associated with achieving maximum profit?

20. A firm has found that the total cost TC (in dollars) of producing x items is given by the equation $TC = \dfrac{1}{3}x^3 - 10x^2 - 800x + 12,000$. Determine the number of units this firm should produce to **minimize** its total production cost.

21. Assume that the firm described in Exercise 20 can sell all units it produces at a fixed price of $325 per unit. Determine how many units this firm should produce to maximize its profit.

22. Find the maximum and minimum values of the function given by

$$y = \begin{cases} -x^2 + 2x + 5 & (0 < x \le 2) \\ x + 1 & (2 < x \le 3) \end{cases}$$

23. Find the maximum and minimum values of the function given by

$$y = \begin{cases} -x^2 + 2x + 5 & (0 \le x \le 2) \\ x^2 - 8x + 16 & (2 < x \le 5) \end{cases}$$

24. A function $y = f(x)$ has a *relative* maximum at $x = c$ if there exists an interval (perhaps very small) centered around $x = c$ such that the maximum value of $f(x)$ over this interval occurs at $x = c$. Determine the relative maxima for the function drawn in Figure 6.4.

25. A function $y = f(x)$ has a *relative minimum* at $x = c$ if there exists an interval (perhaps very small) centered around $x = c$ such that the minimum value of $f(x)$ over this interval occurs at $x = c$. Determine the relative minima for the function drawn in Figure 6.4.

26. Can a relative maximum also be a maximum?

27. Can a relative maximum occur at an end point? Must a maximum be a relative maximum?

28. Theorem 6.1 remains true if the words "maximum" and "minimum" are replaced by "relative maximum" and "relative minimum." In addition, the following result is also valid. *If $f'(c) = 0$ and if the second derivative exists at $x = c$, then $x = c$ is a relative maximum if $f''(c) < 0$, and $x = c$ is a relative minimum if also $f''(c) > 0$*. Use this second–derivative test to show that $f(x) = -x^2 + 2x + 5$ has a relative maximum at $x = 1$.

29. Use the second–derivative test given in Exercise 28 to find all relative maxima and minima for $y = x^3 - 21x^2 + 120x - 100$, $1 \le x \le 13$.

30. If both the first and second derivatives of $f(x)$ are zero at $x = c$, no conclusions can be drawn from the second–derivative test. Show that both derivatives are zero at $x = 0$ for the three functions $y = x^4$, $y = -x^4$ and $y = x^3$. Graph each function and verify that $y = x^4$ has a relative minimum at $x = 0$, $y = -x^4$ has a relative maximum at $x = 0$, and $y = x^3$ has neither at $x = 0$.

6.2 MODELING

A *model* is a representation of a particular situation. A map is a model of a geographical region, a college transcript is a model of a person's academic achievement, and an organization chart is a model of a company's management structure. Many situations, especially dynamic ones that change with time, can be modeled by mathematical equations.

Very few models are ever complete; that is, the model does not reveal every aspect of the situation it represents. A map does not detail weather conditions (unless it is a weather map, in which case it does not detail road construction), a college transcript does not indicate the ease in which grades were achieved, and an organization chart does not detail the personalities of the people filling the slots. Generally, this is of little consequence.

Each model is built for a particular purpose, which usually requires representing only part of a situation. For a person planning a car trip, a road map may be an adequate model. For a person trying to locate the appropriate individual to see in a company, an organization chart is a good model. These models fit the needs of the people using them. This then is the criteria on which models are judged: Does the model adequately fit the needs of the person using the model?

The word "adequate" is important. What is considered adequate by one person may not be adequate to another. Consequently, a good model for one person may be either an inadequate or a bad model for someone else. This is a fact of life. There are no absolutes in modeling. The usefulness of a model is relative to its purpose.

Because many commercial situations are too complex to be modeled in their entirety, simplifying assumptions are usually made that neglect minor contributions. A model of price fluctuations for a given product may not include the effects of a possible strike in the industry supplying raw materials. A model of consumer spending may not include psychological factors. The choice of which factors should be omitted is often subjective and depends on what factors are relevant for the situation being investigated. Omitting a factor is reasonable if the resulting model adequately represents the given situation.

Models are important to management as an aid in decision-making because questions often arise as to which course of action out of many possible ones is the best. By applying each action to the model and observing the effects of these decisions on the model, an optimal decision often can be found. In the succeeding sections, we model specific business situations with mathematical equations. We then use differentiation methods on the models to ascertain an optimal strategy for the business problem at hand.

One simplification we make throughout this chapter is the modeling of discrete variables with continuous ones. Most business quantities can assume only integer values. Automobile sales are given in whole cars; profits are reported to the nearest dollar; the number of employees in an industry is a whole number. A production run of 19.73 cars, a profit of \$195.73869, and 30,198.7 employees are not commercially realistic.

More formally, we say that a quantity or variable is *continuous* if, whenever it assumes two different values, it can assume all numbers between these values.[1] Car sales are not a continuous variable; one can sell two cars or three cars, but not 2.78 cars which is a number between 2 and 3. Profits are not continuous variables. A profit of \$1.95 is realistic, as is a profit of \$1.96, but a profit of \$1.9557 is not realistic. A good example of a continuous variable is time. An order may be placed at 8:07 or at 8:08, and also at 8:07396, although we may have difficulty in measuring it to this accuracy. Variables that are not continuous are called *discrete*. Car sales and profits are *discrete variables*.

Consider again Equation 6.2:

$$P = -x^2 + 120x - 600 \qquad (0 \le x \le 150),$$

which related the profit, P, to the number of units, x, produced for a particular commodity. We assumed that x could be any value between 0 and 150. Realistically, this is not the case because sales are discrete variables that must be whole numbers. One cannot let $x = 49.8$. For the model, however, $x = 49.8$ is a perfectly good number within the domain. It can be substituted into Equation 6.2 with little difficulty.

Equation 6.2 models a discrete variable (number of units sold) by a continuous variable x. The reason for doing this is that differentiation techniques cannot be applied to discrete variables, only to continuous ones. If we wish to optimize a commercial situation with the derivative, we first must convert all discrete independent variables to continuous ones. We consider a model, for example, Equation 6.2, to be valid if it adequately agrees with reality when the dependent variable assumes integer values. Thus, replacing discrete variables with continuous ones is a simple conceptual process made to the final result obtained by the model, should such a replacement make practical sense.

---

[1] The concept of continuous variables is different from that of continuous functions presented in Section 5.3.

6.3 MAXIMIZING SALES PROFIT

Profit maximization is a major goal of for-profit businesses. For companies that sell a physical product, this problem becomes one of determining the quantity of the product that is either produced or purchased, and then subsequently sold, to maximize the profit on the sale of the items. Mathematically solving this problem requires first creating an appropriate model and then, using the optimization techniques developed in the previous section, determining the maximum profit point.

The Model

The mathematical model used for profit maximization is an extremely simple one; it is based on the standard definition of profit as the revenue received in sales minus the cost of either producing or purchasing the items sold. Using the notation developed in Section 2.5 (Break-Even Analysis), we have

$$R = \text{total sales revenue}$$

$$C = \text{total cost incurred in producing or purchasing the items sold}$$

Now, letting

$$P = \text{profit}$$

the mathematical model for profit is

$$P = R - C \qquad \text{(Eq. 6.3)}$$

In this model, P, R, and C are all taken to be functions of x, which is the number of items either produced or purchased and then subsequently sold.

There are two methods employed in using Equation 6.3 to determine the value of x that maximizes profit. Both methods rely on the simplifying assumption, used in this section, that all items, x, produced or purchased, can subsequently be sold.

The most common method is to first derive equations that accurately model R and C. These equations are then substituted into Equation 6.3, the derivative of the resulting single equation for P is taken, and then set to zero. This method is presented first.

A second method is to symbolically take the derivative of the profit equation, as it is written in Equation 6.3, and set it to zero. The resulting equation is then used to determine the relationship between the derivatives of R and C for maximum profit. This second method, quickly and easily, establishes one of the cornerstone theories in economics, that profit maximization is achieved when marginal revenue equals marginal cost. This second method is presented at the end of this section.

Method 1: Using the Profit Equation

The simplest revenue and cost equations are the ones previously presented in Section 2.5.[2] These are:

$$R = px,$$ (Eq. 6.4)

where p the price per unit, and x is the number of items sold, and

$$C = ax + F$$ (Eq. 6.5)

where a is the cost per item, F is the fixed costs, and x is the number of items produced or purchased. Because of the assumption that all items produced or purchased can be sold, x is the same variable in both Equation 6.4 and 6.5.

Example 1 How many items should a manufacturer make and sell to maximize profit if fixed costs are $10,000, each item costs $1.25 to make and sells for $1.50. Due to physical considerations, the manufacturer can produce a maximum of 100,000 units.

Solution We first determine the total revenue and cost equations. From Equation 6.4, the revenue received from selling x items, in dollars, is

$$R = 1.50x$$

Similarly, we use Equation 6.5 to obtain the cost equation. In dollars, the total cost for producing x items is

$$C = 1.25x + 10,000.$$

Because $P = R - C$, we have

$$P = 1.50x - (1.25x + 10,000)$$
$$= 0.25x - 10,000 \quad (0 \le x \le 100,000)$$

Taking the derivative of this function yields

$$P' = 0.25.$$

---

[2] It should be noted that in Section 2.5 the value determined to be the break-even point was based on achieving no profit and no loss (that is $P = 0$). Here we are determining the number of items produced or purchased and subsequently sold to make P a maximum.

Clearly, this derivative can never be zero because there is no value of x that will cause this to happen. Other points where a maximum can occur are the end points $x = 0$ and $x = 100,000$ and points where the first derivative does not exist, of which there are none. Thus, there are only two candidate points, $x = 0$ and $x = 100,000$.

Evaluating the function $P = 0.25x - 10,000$ at the candidate points, we obtain Table 6.6. It follows from this table that the maximum profit is 15,000, which occurs the endpoint of 100,000 items (that is, the largest number of items that can physically be produced). This is reasonable because the profit, P, increases for increasing values of x; the profit is maximized by producing and selling as many items as possible.

TABLE 6.6 Maximum and Minimum Candidate Points.

x	$P = 0.25x - 10,000)$
0	−10,000 <--End Point
100,000	15,000 <--Maximum

The result found in this first example will always be true when the revenue and cost equations are determined by Equations 6.4 and 6.5. The reason is that each additional item produced and sold adds a fixed amount (the difference between the unit price and unit cost) to the profit. However, in practice, these two equations do not always accurately model revenue and cost, and maximum profits do not occur at maximum sales. The reasons for this are

1. The revenue can be a more complicated relationship than given by Equation 6.4. For example, there may be a discount for each item sold.

2. At some stage in an actual operation, the added costs incurred in producing or purchasing additional items become higher due to such things as personnel costs, and increased insurance, equipment, lighting, security, and space requirements. This means that the simple relationship given by Equation 6.5 does not apply.

Examples of these more complicated revenue and cost equations are considered next.

Example 2 Consider that the optimal number of widgets to be produced and sold is to be determined. Due to physical space and machine requirements, a maximum of 100,000 widgets can be manufactured. The units have the following cost and revenue equations, in dollars:

$$C = 0.25x + 10,000$$

and

$$R = (2.25 - .000025x)\, x = 2.25x - 0.000025x^2$$

That is, the cost equation is still determined as a combination of variable and fixed costs, but the unit price, p, has a discount of 0.000025 for each unit sold.

Using Equation 6.3, the profit equation becomes

$$P = R - C = (2.25x - 0.000025x^2) - (0.25x + 10,000)$$

$$= -0.000025x^2 + 2.00x - 10,000 \qquad (0 \le x \le 100,000)$$

Using the differentiation procedure detailed in Section 6.1, we first find

$$\frac{dP}{dx} = -.00005x + 2.00$$

Setting this derivative equal to zero, we calculate

$$-.00005x + 2.00 = 0$$
$$.00005\, x = 2.00$$
$$x = \frac{2.00}{.00005}$$
$$x = 40,000$$

Again, other points where a maximum can occur are the endpoints $x = 0$ and $x = 100,000$ and points where the first derivative does not exist, of which there are none. From Table 6.7, it can be seen that the largest monthly profit is $P = 30,000$ at a production run of 40,000 units, far from capacity.

TABLE 6.7 Maximum and Minimum Candidate Points.

x	$P = 0.000025x^2 + 2x - 10,000)$
0	−10,000 <--End Point
40,000	30,000 <--Maximum
100,000	−60,000 <--End Point

Example 3 A publishers sells a certain soft-cover book for $15 per copy. The revenue and cost equations for this book, in dollars, are

$$R = 15x$$

and

$$C = \frac{x^2}{5,000} + 8x + 12,000$$

Determine the weekly production run that will generate maximum profit. Assume that all books published can be sold, the capacity of the publishing house is 20,000 copies per week, and the publisher has contractual obligations to a distributor for 5,000 copies per week.

Solution The profit equation for this example is

$$P = R - C = 15x - \left(\frac{x^2}{5,000} + 8x + 12,000 \right) = -\frac{x^2}{5,000} + 7x - 12,000$$

To determine a suitable domain, we note that production capacity limits production to 20,000 copies, so $x \le 20,000$. Contractual obligations require a run of at least 5,000 copies, hence $x \ge 5,000$. Together, we have $5,000 \le x \le 20,000$, and the problem becomes one to find values of x that will maximize

$$P = -\frac{x^2}{5,000} + 7x - 12,000 \qquad (5,000 \le x \le 20,000)$$

The first derivative of this function is

$$\frac{dP}{dx} = -\frac{2x}{5,000} + 7$$

Setting this first derivative to zero, we calculate

$$\frac{2x}{5,000} = 7$$

$$2x = 35,000$$

$$x = 17,500$$

Other points where the maximum may exist include the endpoints $x = 5000$ and $x = 20,000$, and points where dP/dx does not exist, of which there are none. It then follows from Table 6.8 that the maximum profit of $P = 49,250$ is realized with a production run of $x = 17,500$ copies per week.

TABLE 6.8 Maximum and Minimum Candidate Points.

x	$P = \dfrac{x^2}{5000} + 7x - 12{,}000$
5,000	−18,000 <--End Point
17,500	49,250 <--Maximum
20,000	−48,000 <--End Point

Method 2: Marginal Revenue Equals Marginal Cost

In this second method, Equation 6.3 is used to determine a general relationship between revenue and cost that produces the same result as obtained using Method 1. This second method is presented as it easily and simply establishes one of the cornerstones of modern economic theory.

Consider Equation 6.3, which, for convenience, is reproduced below

$$P = R - C \qquad \text{(Eq. 6.3)}$$

We know that, in the absence of any constraints on the production or purchase of the items being sold, the maximum profit point is found at the point for which the derivative of P, that is P', is zero. Taking the derivative of Equation 6.3 and setting it equal to zero yields

$$P' = R' - C' = 0$$

Solving this equation, the derivative is zero when

$$R' = C'$$

That is, the maximum profit is achieved when the derivative of the revenue function equals the derivative of the cost function. In economics, the derivative of revenue is referred to as **marginal revenue**, and the derivative of cost is referred to as **marginal cost**. Thus, in economic terms, maximum profit is achieved when marginal revenue equals marginal cost, barring any physical limitations on production (that is, without any restrictions defining endpoints).

From a mathematical perspective, this also provides a second method of maximizing profit. Once a revenue and cost equation has been determined, the derivative of each equation can be taken separately, and these derivatives set equal and solved for x. The value obtained is the same as that produced by setting the derivative of the profit equation equal to zero.

Example 4 Find the optimal value of x for Example 3 by setting the marginal revenue equal to the marginal cost.

Solution The revenue and cost equations, from Example 3 are

$$R = 15x \text{ and } C = \frac{x^2}{5,000} + 8x + 12,000$$

Thus, the marginal revenue, which is the derivative of the revenue function is

$$R' = 15.$$

Similarly, the marginal cost, which is the derivative of the cost function is

$$C' = \frac{2x}{5,000} + 8 .$$

Setting marginal revenue equal to marginal cost, we obtain

$$15 = \frac{2x}{5,000} + 8$$

Solving for x yields

$$x = 17,500.$$

This is the same value that was obtained in Example 3 by setting the first derivative of the profit equation equal to zero. As in Example 3 (see Table 6.8), the endpoints would also have to be considered.

Exercises 6.3

1. A television manufacturer can sell all units produced at $200 per unit. The total cost C (in dollars) in producing x units per week is given by

   $$C = 5,000 + 20x + \tfrac{1}{2}x^2.$$

   a. Determine an expression (model) for profit as a function of x.

   b. Determine an appropriate domain for x if the weekly production capacity is limited to 300 units.

   c. Determine the maximum weekly profit.

2. Redo Exercise 1 if the weekly production capacity is only 150 units.

3. A company can sell all units of a particular product. The cost equation (in dollars) of producing x units is given by

    $$C = \frac{x^2}{4,000} - 5x + 50,000.$$

    The unit price is fixed at $10 per item.
    a. Determine an expression for the profit as a function of x,

    b. Determine an appropriate domain for x if the production capacity is limited to 50,000 units.

    c. Find the maximum profit.

4. Redo Exercise 3 if, in addition, the manufacturer has contractual obligations for 40,000 units per week.

5. Assume that the manufacturer described in Exercise 3 is interested in minimizing costs rather than maximizing profits. How many units should be produced to achieve this objective?

6. A refrigerator manufacturer can sell all the refrigerators it can produce. The total cost of producing x refrigerators per week is given by the equation $C = 300x + 900$. The unit price, p, is related to the number of refrigerators sold by the equation $p = 700 - 2x$.

    a. Determine an equation for the revenue equation.

    b. Determine an equation for the profit.

    c. Determine an appropriate domain if the production capacity is 100 units per week.

    d. How many units should be produced in a week to maximize profits?

7. Redo Exercise 6 if the production capacity is only 40 refrigerators per week.

8. A clothing manufacturer can sell all the suits produced each month up to a limit of 500. The unit cost per suit is $150, and the fixed costs are $15,000. The unit price, in dollars, is given as $UP = 200 - 0.25x$. Determine the maximum profit the company can realize each month.

9. Determine the marginal revenue if
    a. $R = 15x$
    b. $R = 200x - (1/4)x^2$.
    c. $R = 2.25x - 0.000025x^2$.

10. Evaluate the marginal revenue found in Exercise 9c for both $x = 30,000$ and $x = 50,000$.

11. Determine the marginal cost if

    a. $C = (x^2/5,000) + 8x + 12,000$

    b. $C = 50x + 5,000$

    c. $C = 10,000 + 0.25x$.

12. Evaluate the marginal cost found in Exercise 11c. for both $x = 30,000$ and $x = 50,000$.

13. Redo Example 2 by setting the marginal revenue equal to the marginal cost.

14. Find the optimal value of the quantity x in Exercise 3 by setting the marginal revenue equal to the marginal cost.

15. Find the optimal value of the refrigerators in Exercise 6 by setting the marginal revenue equal to the marginal cost.

6.4   MINIMIZING INVENTORY COSTS

Optimization techniques are used in inventory problems for minimizing the total of storage and ordering costs while simultaneously ensuring that enough items are on hand to meet current demand. In this section, we present a simplified inventory model and then determine the optimum number of items to stock to minimize the costs of ordering and storing inventory items. Inherently, we wish to avoid both over-stocking, with its resulting increase in storage costs, and under-stocking, with its resulting loss in sales. Table 6.9 presents the parameters that are assumed known; they are used to determine the optimum amount of items to be ordered with each order placed throughout the year.

TABLE 6.9 Known Inventory Parameters.

Notation	Meaning	Comments
D	Annual demand for all of the items	A known amount.
m	Cost of placing a single order	A known dollar amount.
k	Cost of storing one item for one year.	Either given or calculated as the cost of financing times the cost of a single item.
f	Cost of financing, as a percent	Used to calculate k, if k is not given.
c	Cost of a single item	Used to calculate k, if k is not given.
t	Lead time	The delivery time, in days, between when an order is place and when it is received.

Table 6.10 lists the quantities that we will be calculating, with the last item in the table being the critical one – the EOQ.

TABLE 6.10 Calculated Inventory Parameters.

Notation	Meaning	Formula
x	Number of items to be ordered with each order	The unknown variable
N	Number of orders placed in a year	$N = \left(\frac{D}{x}\right)$
AI	Average items in inventory	$AI = \left(\frac{x}{2}\right)$
k	Annual storage cost for one item	$k = (f)(c)$, or it is given directly
TOC	Total Annual Ordering Cost	$TOC = (m)(N) = m\left(\frac{D}{x}\right)$
TSC	Total Annual Storage Cost	$TSC = k\left(\frac{x}{2}\right)$
TC	Total Annual Cost	$TC = TOC + TSC$
EOQ	Economic Order Quantity (Optimum Order Size)	$EOQ = x = \sqrt{\dfrac{2mD}{k}}$

The Model

In this problem, the goal is to determine the optimum order size, referred to as the ***Economic Order Quantity*** (**EOQ**), that minimizes the total yearly inventory costs while maintaining sufficient inventory at all times to meet the daily demand for the inventoried items.

In addition to the quantities that are assumed known (see Table 6.9), two additional assumptions are made. These are:

1. The demand for items is uniform throughout the year. That is, as many items are sold during the first day as are sold during the 200th day, and as many items are sold during the fourth week as are sold during the seventeenth week.

2. Inventory is reordered at equal time intervals and in equal lots. For example, 700 cases every 2 weeks.

These two assumptions imply a depletion of inventory, as shown in Figure 6.5

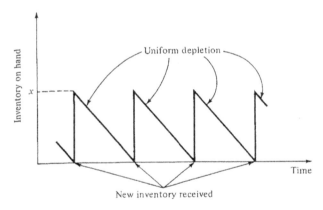

FIGURE 6.5 Uniform inventory depletion.

In this model, it is the total yearly cost, TC, that we want to minimize. This total cost is simply the sum of the total yearly ordering costs, TOC, and the total yearly storage cost, TSC, for the items. Mathematically,

$$TC = TOC + TSC \qquad \text{(Eq. 6.6)}$$

The total yearly ordering cost, TOC, is determined as:
$$TOC = (\text{Cost of placing one order})(\text{Number of orders placed in a year})$$

Thus,

$$TOC = m\left(\frac{D}{x}\right) \qquad (Eq.6.7)$$

The second cost needed to minimize total cost is the yearly storage costs, *TSC*. This cost is determined as

TSC = (Cost of storing one item for a year) (Average items in inventory)

Thus,

$$TSC = k\left(\frac{x}{2}\right) \qquad (Eq.6.8)$$

Substituting for *TOC* and *TSC* into Equation 6.6 yields

$$TC = m(D/x) + k(x/2) \qquad (1 \leq x \leq D)$$

The domain of $1 \leq x \leq D$ is determined as follows: Realistically, no fewer than one item can be ordered at a time, and the maximum number of items that can be ordered is *D*, the amount of items needed in a year. Note that ordering *D* items at one time means that only one order is placed in a year).

How Much to Order

Before taking the derivative of the total cost function and setting it equal to zero, we first re-write it as follows

$$TC = mDx^{-1} + (k/2)x \qquad (1 \leq x \leq D) \qquad (Eq. 6.9)$$

Taking the derivative of Equation 6.9 and setting it equal to 0, we have,

$$d(TC)/dx = -mDx^{-2} + (k/2) = 0$$

Solving for *x* results in

$$x^2 = 2mD/k$$

or

$$x = \pm\sqrt{\frac{2mD}{k}}$$

Because $x = -\sqrt{2mD/k}$ is negative, and therefore not in the domain defined by Equation 6.9, we disregard it. Although the first derivative $d(TC)/dx$ does *not* exist at $x = 0$ (because the quantity $(mD)/x$ is undefined at this point), this point is also not in the domain and is disregarded. Thus, a minimum value of total cost can occur only at the endpoints $x = 1$ and $x = D$, or at

$$x = \sqrt{\frac{2mD}{k}}$$

This value of x is referred to as both the *optimum order size* and the *Economic Order Quantity, EOQ*. Thus,

$$EOQ = \sqrt{\frac{2mD}{k}} \qquad \text{(Eq.6.10)}$$

Example 1 A distributor estimates annual demand for television sets to be 1,000 units over the next year. The cost of placing a single order is $10, and storage costs per unit per year are $8. To minimize total inventory costs, determine

a. the optimum order size, and

b. the number of orders that will be placed in a year.

Solution For this problem $D = 1,000$, $m = 10$, and $k = 8$, where all monetary figures have been expressed as dollars. With these values,

a. From Equation 6.10, the Economic Order Quantity is:

$$EOQ = \sqrt{\frac{2mD}{k}} = \sqrt{\frac{2(10)(1,000)}{8}} = \sqrt{2,500} = 50$$

Thus, the minimum occurs either at $x = 50$ or at the end points $x = 1$ and $x = 1,000$.[3] Substituting these values of x into the equation for *TC*, we generate Table 6.11, from which it follows that the optimal order size is $x = 50$, which is the Economic Order Quantity, *EOQ*.

TABLE 6.11 Optimization Candidates.

x	$TC = m\left(\dfrac{D}{x}\right) + k\left(\dfrac{x}{2}\right) = 10\left(\dfrac{1,000}{x}\right) + 8\left(\dfrac{x}{2}\right)$
1	$10,004.
50	$400. <--Minimum
1000	$4,010.

---

[3] The minimum will always occur at the EOQ. The only time order size is limited occurs when the EOQ is larger than the maximum amount that can be ordered, due to physical restrictions on the order.

b. The total number of orders placed in a year, N, is determined from the equation $N = D/x = 1{,}000/50 = 20$ (see Table 6.10). That is, one thousand units are required over the year, and each order is for 50 items; hence, the distributor must place 1000/50 = 20 orders.

Example 2 A soda distributor's annual demand for cases of soda is 2,000 a year. The cost of placing a single order is $9.66, and the storage costs per case per year is $1.40. To minimize total inventory costs, determine:

a. the optimum order size, and

b. the number of orders that will be placed in a year.

Solution Here $D = 2{,}000$ $m = 9.66$ and $k = 1.40$, where all monetary figures have been expressed as dollars. With these values,

a. From Equation 6.10, the Economic Order Quantity, rounded to the nearest single unit, is:

$$EOQ = \sqrt{\frac{2mD}{k}} = \sqrt{\frac{2(9.66)(2{,}000)}{1.40}} = \sqrt{27{,}600} = 166$$

Unless there is a restriction that limits an order to below this amount, the EOQ will produce the minimum cost. That is, the endpoints need not be checked because they do not restrict an order to be less than 166 cases, or at the endpoints of 1 and 2,000. Substituting the value of 166 into the total cost equation yields:

$$TC = m\left(\frac{D}{x}\right) + k\left(\frac{x}{2}\right),$$

which yields a total cost of:

$$TC = 9.66\left(\frac{2{,}000}{166}\right) + 1.40\left(\frac{166}{2}\right) = \$232.58$$

b. The total number of orders placed in a year, N, is determined as $D/x = 2000/166 = 12.04$. Thus, rounded to the nearest unit, the minimum inventory cost is achieved by placing 12 orders throughout the year.

When to Order

In addition to determining *how many items* to order with each order, and *how many orders* will be needed in a year, it is also important to know *when* to place an order. This determination depends on the *lead time*, which is the estimated time between when an order is placed and when it is received.

Take a look at Figure 6.5 again, reproduced below as Figure 6.6 for convenience.

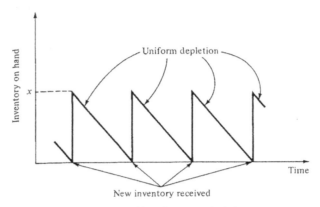

FIGURE 6.6 Uniform inventory depletion.

The model represented by Figure 6.6 shows that new inventory is received exactly when the existing inventory goes to zero. In practice, this is never true because daily demand is never exactly uniform. To compensate for this, orders are typically placed when inventory falls below a pre-determined point and based on the fact that there is a delay between when an order is placed and when it is received.

The point at which an order is placed is referred to as the *reorder point, RP,* and it is calculated as daily demand, d, times the lead time, t. Thus,

$$RP = (d)(t) \qquad \text{(Eq. 6.11)}$$

where the daily demand, d, is calculated as

$$d = \frac{D}{Working\,days\,in\,a\,year}$$

Because the reorder point also assumes an exact daily demand, in practice, the reorder point is raised slightly from the number determined in Equation 6.11. Doing this adjusts for any spikes that can occur in the daily demand before a new order is received in inventory.

The last quantity of interest is the cycle time, T, which is just the time between when orders are placed. The formula for cycle time is

$$T = \frac{Working\,days\,in\,a\,year}{Number\,of\,orders\,placed\,in\,a\,year}$$

Table 6.12 summarizes these additional formulas.

TABLE 6.12 Additional Calculated Inventory Parameters.

Notation	Meaning	Formula
d	Daily demand	$d = \dfrac{D}{Working\,days\,in\,a\,year}$
RP	Reorder Point	$RP = (d)(t)$
T	Cycle time (the time between the placement of when orders.)	$T = \dfrac{Working\,days\,in\,a\,year}{Number\,of\,orders\,placed\,in\,a\,year}$

Exercises 6.4

1. The cost of placing an order is $10, and the cost of storing one item for one year is $3. Determine the optimum order size if 540 items are required during a year.

2. The cost of placing an order is $1, and the cost of storing one item for one year is $25. Determine the economic order quantity if 31,250 items are required in a year.

3. The cost of placing an order is $4.80, and the storage cost per unit for one month is $2. Find the optimum order size if 1,000 units are needed each year. (Hint: convert all parameters to yearly values.)

4. For the Economic Order Size determined in Exercise 3, determine:

   a. the total reorder costs over the course of a year.

   b. the total storage costs over the course of a year.

   c. the total costs incurred over the course of a year.

5. The cost of placing an order is $9.60. One thousand units are required during a year, and the cost of carrying one item for 1 year is $1.92.

   a. Determine the economic order quantity.

   b. Find the number of orders that will be placed during a year.

6. For the situation described in Exercise 5 and the results obtained in parts a. and b., construct a figure similar to Figure 6.5.

7. Assume that Figure 6.7 adequately describes the inventory situation of a medium sized appliance dealer.

 a. Determine the economic order quantity for the dealer.

 b. Find the total number of orders placed during a year.

 c. Determine the total number of appliances sold during a year.

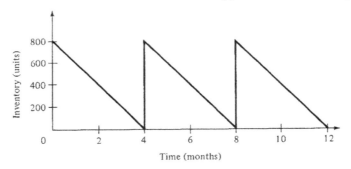

FIGURE 6.7

8. Determine the reorder point and cycle time for Exercise 1, assuming the lead time is 4 days.

9. Determine the reorder point and cycle time for Exercise 2, assuming the lead time is 2 days.

10. Determine the reorder point and cycle time for Exercise 3, assuming the lead time is 3 days.

11. Using the second derivative test described in Exercise 28 in Section 6.1, show that the economic order quantity $x = \sqrt{2mD/k}$ is a relative minimum for Equation 6.9. From this, conclude that whenever the economic order quantity lies in the domain $1 \le x \le D$ it is also the minimum.

12. Equation 6.9 presupposes a linear relationship between the order size x and the individual order cost, TOC. Derive an expression for the total storage cost TC if the relationship between x and TOC is quadratic; that is, find a formula for TC if $TOC = mx + px^2$. What is the economic order quantity in this case?

6.5 ECONOMETRICS

Econometrics is the branch of economics that uses mathematical methods to model economic systems. Although econometrics relies heavily on statistical methods, an important but simple application of the derivative as a rate of change is also used. By applying differentiation techniques to a simplified model of a national economy, for example, we can gain interesting insights into the effect of an increase in overall business investment on the total economy. These insights are the beginnings of Keynesian Economic theory.

Our first assumption is that the economy consists of only consumers (i.e., individuals) and businesses, ignoring the effects of government and foreign sectors. Such a model approximates conditions in the United States during the previous century. In the exercises, we refine this model to include the government.

Our second assumption is that businesses do not save money. All money received by the business is distributed to banks to repay past loans or to stockholders in the form of dividends. Banks, being themselves businesses, return their receivables to individuals as interest payments. Therefore, all money received by businesses is eventually distributed to individuals as dividends or interest. Our second assumption can be restated as, "All money spent by consumers and businesses is received only by consumers."

Each dollar received by consumers is either spent or saved. Because, by our second assumption, banks cannot save money, they are assumed to lend all available funds to businesses to use for commercial expenditures. Businesses borrow money from banks for investment. As these investments produce money, the profits are used to both repay loans from the banks and as share-holder dividends. Consumers receive money from other consumers, or from banks in the form of interest payments, or from businesses in the form of dividends. This money is either re-spent directly or saved. Saved money is lent to business for investment, and the cycle begins again.

In summary, we are making the following simplifying assumptions:

1. The economy consists of only consumers and businesses.

2. All money is received only by consumers.

3. Businesses borrow from consumer savings to obtain funds for business expenditures.

We are interested in the question, What are the effects of a change in business investment on the total economy?

Before attempting an answer, we must determine how we will measure the total economy. A reasonable measure is the *gross national product* which is simply the total amount of money spent within the economy. We denote this quantity by T. If we let B denote the amount of money spent by businesses, the question of interest can be restated: What is the change in T with respect to a change in B? Or, mathematically, what is dT/dB?

The Model

To calculate this derivative, we first must express T as a function of B. That is, we must model our economy in such a way that we can ultimately obtain one equation for T in terms of B. We already have

$$T = \text{total expenditures within the economy}$$

and

$$B = \text{total business expenditures.}$$

In addition, we now define

$$C = \text{total consumer expenditures.}$$

It then follows directly from the first assumption that

$$T = C + B \tag{Eq. 6.12}$$

Equation 6.12 is a model of the economy, but it is not sufficient for our purpose. We need an equation for T strictly in terms of B (so we can differentiate), whereas Equation 6.12 relates T to both B and C. More information is needed.

From all available data, it appears that consumer spending C can be decomposed into two parts, necessities, and luxuries. Necessities are the absolute basic necessities, and luxuries are everything else. For example, a family may have to spend \$100 per week for food to survive but, if more funds were available, they might spend the additional funds for additional food or entertainment. This additional amount would be classified as a luxury. Expenditures for basic necessities, which we denote as C_o, are reasonably fixed. Expenditures for luxuries are a function of the amount of money available; the more money available, the more money spent on luxuries.

Because all money spent by consumers and businesses T is received by consumers [2nd assumption], we see that expenditure for luxuries is a function of T. Furthermore, if we assume that this portion of consumer spending is proportional to the amount of money available, we have

$$C = mT + C_o \qquad \text{(Eq. 6.13)}[4]$$

where m denotes the fraction of total income spent on luxuries.

To simplify the problem even further, we assume we are modeling an affluent society in which expenditures for basic necessities are much less than expenditures for luxuries. In such a case, we can replace Equation 6.13 with

$$C = mT \qquad \text{(Eq. 6.14)}$$

with little error. In economic theory, the term m is known as the *marginal propensity to consume*. Substituting Equation 6.14 into Equation 6.12 and rearranging, we find

$$T = mT + B.$$

Solving for T yields,

$$T = \left(\frac{1}{1-m}\right)B \qquad \text{(Eq.6.15)}$$

Equation 6.15 is the model we were seeking. Differentiating it with respect to B, we obtain the desired rate of change

$$\frac{dT}{dB} = \left(\frac{1}{1-m}\right) \qquad \text{(Eq.6.16)}$$

Equation 6.15 is the equation of a straight line, and Equation 6.16 is its slope. It follows directly from our knowledge of slopes (Section 2.4) that an increase in business expenditures B by an amount Q forces a corresponding increase in T by $1/(1\text{-}m)$ times Q. In economic theory, the term $1/(1-m)$ derived in Equation 6.16 is known as the **multiplier**.

Example 1 Determine the effect of a $10,000 increase in business investment on the total national income if individuals as a group spend 75% of the national income.

---

[4] This was one of the key assumptions made by Lord Keynes in his general theory of employment, interest, and money.

Solution Because all monies spent within the economy are received by consumers [assumption 2], total national income is the same as total expenditures T. A $10,000 increase in business investment is the same as an increase in business expenditures B of $10,000. From Equation 6.16, the change in T with respect to B is $1/(1 - m)$. Here $m = 0.75$, hence

$$\frac{1}{1-m} = \frac{1}{1-0.75} = \frac{1}{0.25} = 4$$

That is, T increases 4 units with every one-unit change in B. A $10,000 change in B results in a theoretical change of $40,000 in T.

At first glance, this result is startling. Example 1 indicates that the total income for society as a whole can be increased by $4 with only a $1 increase in business investment, assuming $m = 0.75$. A little thought reveals why this is so.

After the initial expenditure of $1, the recipient, a consumer, receives a dollar as income. He or she spends 75¢ (75% of $1) and saves the rest. A second consumer receives the 75¢ spent by the first consumer. He or she, in turn, spends 56¢ (75% of 75¢) and saves the rest. A third consumer receives the 56¢ spent by the second consumer and continues the cycle. As this income is received and re-spent, the net effect will be an increase of $4 in the total income of society.

Of course, these results are valid only to the extent that our model adequately represents a given society. It is unlikely that any economy, including the United States economy during the previous century, can be adequately modeled by Equations 6.12 and 6.14. The model does not include the time period required to re-spend increased income, the economic state of the people receiving this income, or the ability of society to produce the extra goods demanded by consumers having additional money. Unemployment and inflation are not parts of our model. Population growth, food supplies, and available resources have also been neglected. Nevertheless, the construction presented here is typical of that applied to other more sophisticated models.

Exercises 6.5

1. Determine the increase in the total national income of a society modeled by Equation 6.12 and Equation 6.14 if $m = 0.6$ and business investment is increased by $25 million.

2. Determine the decrease in the total national income of a society modeled by Equation 6.12 and Equation 6.14 if $m = 0.7$ and business investment is decreased by $10 million.

3. Redo Exercise 1 for a society modeled by Equations 6.12 and 6.13.

4. As was noted in the text, the constant m used in Equation 6.14 is referred to as the ***marginal propensity to consume***. The term $1/(1 - m)$ derived in Equation 6.16 is known as the ***multiplier***. Determine the value of the multiplier if the marginal propensity to consume is 0.8. Why do you think the term "multiplier" is used?

5. Determine the net effect on the income of a society after a $1,000 increase in business expenditures has been cycled through five people if $m = 0.6$.

6. Consider a three–sector economy consisting of individual consumers, businesses, and governments each making expenditures denoted as C, B, and G, respectively. Assume that the following three equations adequately describe the relationship between these quantities:

$$T = C + B + G$$
$$C = mT + C_o$$
$$B = B_o,$$

where B_o denotes a constant business expenditure and the other symbols are as previously defined in the text. For this model, determine the rate of change in total expenditures with respect to a change in government expenditures.

7. Determine the decrease in the total national income of a society modeled by Equations 6.12 and 6.14 if $m = 0.75$ and business investment is decreased by $8 million.

8. Redo Problem 7 for a society modeled by Equations 6.12 and 6.13.

6.6   SUMMARY OF KEY POINTS

Key Terms

- Critical point
- Cycle time
- EOQ
- Economic order quantity
- Global maximum
- Global minimum
- Lead time
- Optimization

- Relative maximum
- Relative minimum
- Reordering cost
- Storage costs

Key Concepts

- Optimization is the process of maximizing or minimizing a quantity. One of the key methods of determining optimization points is to take the derivative of the desired function, and determine the values that make the derivative zero. The resulting values are candidates for the maximum and minimum points.
- Because all real-world processes are constrained by finite resources, each input in an optimization problem is bounded by a defined limit.
- A domain should include its endpoints when a function models a real-world process.
- A function may attain its global maximum or global minimum at two or more values, one of which may be an endpoint.
- Optimization techniques are used in inventory theory to minimize storage costs while ensuring that sufficient items are available to meet demand.

CHAPTER 7

CURVE FITTING AND TREND LINES

In this Chapter

7.1 Constant Curve Fit
7.2 Linear Least-Squares Trend Lines
7.3 Creating Trend Lines Using Excel
7.4 Selecting an Appropriate Curve using R^2
7.5 Summary of Key Points

Throughout this book, we have used equations to model and represent various business situations without mentioning, for the most part, how the equations were obtained. In practice, such equations are not immediately available to the user but must be derived. The derivations are generally one of two types, theoretical or empirical.

In the theoretical approach, known principles are used to generate the equations. Indeed, this was the approach taken in Chapter 6 to mathematically model the problems presented there. A specific example is given at the beginning of Section 6.3. There we used the accepted principle that profit, P, is the total revenue received from all sales, R, minus the total cost, C.

In the empirical approach, one uses past data to generate the equations. Although such methods are no better than the accuracy of the data, they are often the only ones available, especially if there are no theoretical results that apply. As an example, a certain business, knowing that sales volume depends on advertising expenditures, may want to know the relationship between these quantities. Does volume increase linearly (as a straight line) with advertising expenditures or perhaps exponentially? Obviously, there is no theoretical principle that can answer this question because the answer differs from product to product. In this chapter, we present an introduction to curve fitting, which is one method of extracting meaningful mathematical equations from a set of data points. Also presented is how spreadsheet programs, such as Excel®, can be used to both graphically display and provide these mathematical equations.

7.1   CONSTANT CURVE FIT

The simplest curve fit occurs when the data are relatively constant. For this case, a horizontal straight line (see Section 2.3) may represent a good approximation to the given situation.

Definition 7.1 Let a denote the average value of a set of data. The line $y = a$ is the average-value, straight-line fit for these data points.

Example 1 The yearly gross sales of a manufacturing firm for the past decade are plotted in Figure 7.1. Determine the average-value, straight-line fit for these data.

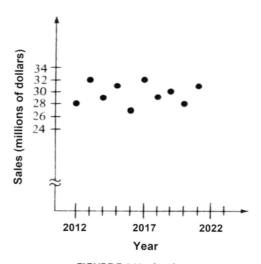

FIGURE 7.1 Yearly sales.

Solution For convenience, we first tabulate the given data points in Table 7.1. Because the data are relatively constant, a straight-line fit is a reasonable approximation. Here the average yearly sales are

$$a = \frac{28 + 32 + 29 + 31 + 27 + 32 + 29 + 30 + 28 + 31}{10} = 29.7$$

TABLE 7.1 Yearly Sales.

Year	Sales (millions)	Year	Sales (millions)
2012	28	2017	32
2013	32	2018	29
2014	29	2019	30
2015	31	2020	28
2016	27	2021	31

The average-value, the straight-line fit, is $y = 29.7$, which is drawn in Figure 7.2.

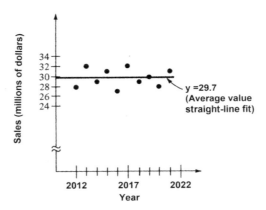

FIGURE 7.2 Creating an average value Straight-line fit.

Obviously, if the data are not reasonably constant, the average-value method is a poor approximation that can lead to erroneous conclusions. We would have little confidence in a 2023 projection based on the average-value, straight-line fit for the data given in Figure 7.3.

A useful modification to an average-value, straight-line fit is the concept of moving averages. Here several averages are calculated for different time periods, resulting in a set of averages which "move with the data."

Example 2 Determine consecutive 5-year moving averages for the data previously listed in Table 7.1.

Solution The first 5-year average a_1, includes data for 2012 through 2016, is

$$a_1 = \frac{28 + 32 + 29 + 31 + 27}{5} = 29.4$$

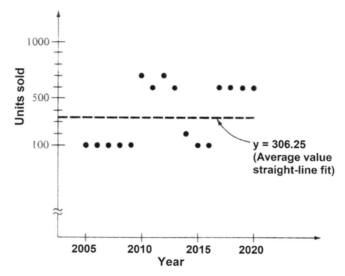

FIGURE 7.3 Creating an average value straight-line fit.

The second 5-year average for the years 2013 through 2017 is

$$a_2 = \frac{32 + 29 + 31 + 27 + 32}{5} = 30.2$$

Using the data for the years 2014 through 2018, we compute the third 5-year moving -average as

$$a_3 = \frac{29 + 31 + 27 + 32 + 29}{5} = 29.6$$

Continuing in this manner, we also find that $a_4 = 29.8$, $a_5 = 29.2$, and $a_6 = 30.0$.

The arithmetic for computing consecutive moving averages can be simplified if we note that each average differs from its predecessor by the addition and deletion of two pieces of data. In Example 2, a_2 differed from a_1 by the addition of the 2017 data and the deletion of the 2012 data. Similarly, a_3 differed from a_2 by the addition of the 2018 data and the deletion of the 2013 data. In general, each new average can be calculated from the previous average by adding the difference between the new data point and the oldest data point divided by the time span under consideration. Thus, in Example 2, a_2 can be determined from a_1 as

$$a_2 = a_1 + \left(\frac{32 - 28}{5} \right)$$

and

$$a_3 = a_2 + \left(\frac{29 - 32}{5} \right)$$

Example 3 Compute and graph consecutive 4-year moving averages for the years 2014 through 2020, inclusive, for the data given in Figure 7.3.

Solution The first 4-year moving average that is being calculated now starts at 2014 and ends at 2017. Reading the appropriate points from the graph, we find

$$a_1 = \frac{200 + 100 + 100 + 600}{4} = 250.$$

Then,

$$a_2 = a_1 + \left(\frac{600 - 200}{4} \right) = 250 + 100 = 350,$$

$$a_3 = a_2 + \left(\frac{600 - 100}{4} \right) = 350 + 125 = 475,$$

and

$$a_4 = a_3 + \left(\frac{600 - 100}{4} \right) = 475 + 125 = 600.$$

These averages are drawn in Figure 7.4

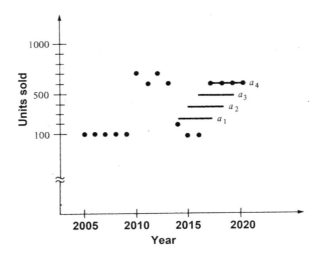

FIGURE 7.4 Creating 4-year moving averages.

Exercises 7.1

1. Find the average-value, straight-line fit for the data given in Table 7.2.

TABLE 7.2

Price ($)	1	2	3	4	5	6	7	8	9	10
Demand (in 100s)	12	13	9	8	11	10	9	13	8	10

2. Compute consecutive 5-year moving averages for the data given in Table 7.2.

3. The gross sales for a small furniture store are given in Table 7.3 for a 9-year period. Find the average-value, straight-line fit for these data.

TABLE 7.3

Year	2013	2014	2015	2016	2017	2018	2019	2020	2021
Sales ($1,000,000)	28	29	32	327	26	31	28	27	30

4. The Spencer Food Company is considering an increase in its advertising budget to bolster sales of its Kitty-Kat high-protein cat food, a product that has particular appeal to upper-income cat owners. As a manager of the company, you are asked for a preliminary opinion on the advisability of such an increase. What is your initial reaction based on the information listed in Table 7.4?

TABLE 7.4

Year	2013	2014	2015	2016	2017	2018	2019	2020	2021
Advertising ($1,000)	140	150	160	170	180	160	160	170	170
Sales ($1,000,000)	28	29	32	26	26	27	28	27	30

7.2 LINEAR LEAST-SQUARES TREND LINES

Empirically obtained data on supply, demand, sales, cost, and other commercial quantities rarely are represented adequately by average-value straight lines. Many times, however, such data can be modeled by the more general straight line $y = mx + b$, with $m \neq 0$.

If the data consist of only two points, we can use the methods developed in Section 2.3 (see Example 7) to fit a straight line through them. If more than two data points are given, one of two situations can occur. First, all the data points can lie on the same straight line. In such a case, which almost never occurs in practice, we simply pick two of the points and construct a straight line through them, as before.

The more common situation involves a set of data that do not lie on any straight line but which, nonetheless, seem to be adequately represented by such a curve. A case in point involves the data plotted in Figure 7.5. Although a straight line appears to be a reasonable approximation to the data, no one line of the form $y = mx + b$ contains all the given points. Therefore we seek the straight line that best fits the data.

Any straight-line approximation has one y-value on the line for each value of x. This y-value may or may not agree with the given data. Thus, for the values of x at which data are available, we generally have two values of y: one value from the data and a second value from the straight-line approximation to the data.

This situation is illustrated in Figure 7.6. The error at each x- is simply the difference between the y-value of the data point and the y-value obtained from the straight-line approximation. We designate this error as $e(x)$.

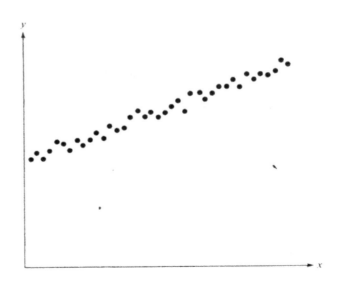

FIGURE 7.5 Data indicating a straight-line trend line.

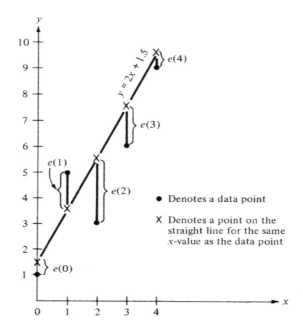

FIGURE 7.6 Determining trend line error.

Example 1 Calculate the errors made in approximating the data shown in Figure 7.6 by the line $y = 2x + 1.5$.

Solution The line and the given data points are plotted in Figure 7.6. There are errors at $\underline{x} = 0$, $x = 1$, $x = 2$, $x = 3$, and $x = 4$. Evaluating the equation $y = 2x + 1.5$ at these values of x, we compute Table 7.5.

TABLE 7.5

Given data		Evaluated from $y = 2x + 1.5$
x	y	y
0	1	1.5
1	2	3.5
2	3	5.5
3	6	7.5
4	9	9.5

It now follows that

$$e(0) = 1 - 1.5 = -0.5$$
$$e(1) = 2 - 3.5 = -1.5$$
$$e(2) = 3 - 5.5 = -2.5$$
$$e(3) = 6 - 7.5 = -1.5$$

and

$$e(4) = 9 - 9.5 = -0.5.$$

Note that these errors could have been read directly from the graph.

We can extend this concept of errors to the more general situation involving N data points. Let (x_1, y_1), (x_2, y_2), (x_3, y_3), ..., (x_N, y_N) be a set of N data points for a particular situation. Any straight-line approximation to this data generates errors $e(x_1)$, $e(x_2)$, $e(x_3)$, ... , $e(x_N)$ which individually can be positive, negative, or zero. The latter case occurs when the approximation agrees with the data at a particular point. We define the overall error as follows.

Definition 7.2 The least-squares error, E, is the sum of the squares of the individual errors. That is,

$$E = \left[e(x_1)\right]^2 + \left[e(x_2)\right]^2 + \left[e(x_3)\right]^2 + \ldots \left[e(x_N)\right]^2 \qquad \text{(Eq. 7.1)}$$

Using this definition, the only way the total error, E, can be zero is for each of the individual errors to be zero. The reason for this is that each term in Equation 7.1 is squared, which ensures that an equal number of positive and negative individual errors cannot sum to zero.

Example 2 Compute the least-squares error for the approximation used in Example 1.

Solution
$$E = \left[e(0)\right]^2 + \left[e(1)\right]^2 + \left[e(2)\right]^2 + \left[e(3)\right]^2 + \left[e(4)\right]^2$$
$$= (-0.5)^2 + (1.5)^2 + (-2.5)^2 + (-1.5)^2 + (-0.5)^2$$
$$= 0.25 + 2.25 + 6.25 + 2.25 + 0.25 = 11.25.$$

Definition 7.3 The least-squares straight line is the line that minimizes the least-squares error.

The least-squares straight line is given by the equation $y = mx + b$, where m and b simultaneously satisfy the two equations

$$bN + m\sum_{i=1}^{N} x_i = \sum_{i=1}^{N} y_i \qquad \text{(Eq. 7.2)}$$

and

$$b\sum_{i-1}^{N} x_i + m\sum_{i=1}^{N}\left(x_i\right)^2 = \sum_{i=1}^{n} x_i y_i \qquad \text{(Eq. 7.3)}$$

Here x_i and y_i denote the values of the ith data point and N is the total number of data points being considered. We also have used the sigma notation introduced in Section 1.7.

Example 3 Find the least-squares straight line for the data listed in Table 7.6.

TABLE 7.6 Determining a Least-Squares Straight-Line for the following Data.

x	0	1	2	3	4
y	1	5	3	6	9

Solution A good procedure for calculating the least-squares straight line is to first construct an expanded table similar to Table 7.7.

TABLE 7.7

	x_i	y_i	$\left(x_i\right)^2$	$x_i y_i$
	0	1	0	0
	1	5	1	5
	2	3	4	6
	3	6	9	18
	4	9	16	36
Sums:	10	24	30	65

Once this table is constructed, the following sums can be directly obtained as:

$$\sum_{i=1}^{5} x_i = 10$$

$$\sum_{i=1}^{5} y_i = 24$$

$$\sum_{i=1}^{5}\left(x_i\right)^2 = 30$$

and

$$\sum_{i=1}^{5} x_i y_i = 65$$

Substituting the appropriate sums into Equations 7.2 and 7.3, with $N = 5$ (because there are 5 data points), we have

$$5b + 10m = 24$$

and

$$10b + 30m = 65.$$

Solving these two equations simultaneously for m and b, we obtain $m = 1.7$ and $b = 1.4$. Thus, the equation of the least-squares line is

$$y = 1.7x + 1.4$$

The graph of this line, along with the original data is shown in Figure 7.7.

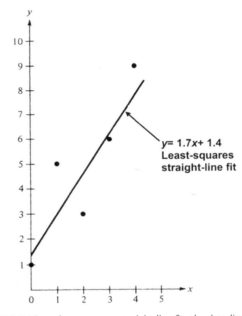

FIGURE 7.7 Linear least-squares straight-line for the data listed in Table 7.7.

7.3 CREATING TREND LINES USING EXCEL

In business, economics, and science, a line used to fit empirically obtained data is referred to as a *trend line*. A trend line can be linear, quadratic, exponential, or any other line that best fits the data. Mathematically, a best fit is accomplished by minimizing the total squared error between the trend line and the actual data. The dotted line shown in Figure 7.8 is an example of a straight line trend line.

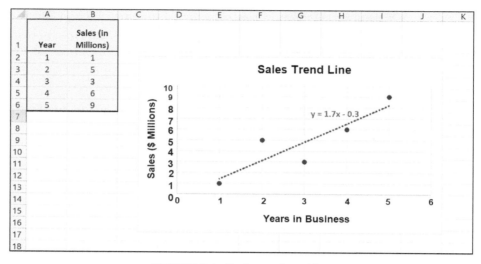

FIGURE 7.8 An Excel created trend line.

In constructing a trend line, the first step is to create a scatter diagram. A scatter diagram is a graph containing multiple data points, with no connecting lines between the points[1]. Figure 7.9 illustrates two examples of scatter diagrams. Trend lines are graphed on the resulting scatter diagram.

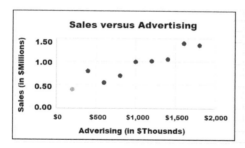
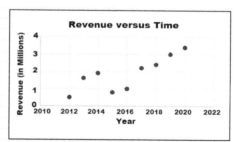

FIGURE 7.9 Examples of scatter diagrams.

Creating a Scatter Diagram

Creating a scatter diagram requires that the data is first entered into either consecutive columns, as shown in the upper-left corner of Figure 7.10, or consecutive rows. The data is then highlighted, as shown as Step 1 in the figure, and the remaining steps listed in the figure are then completed.

---

[1] When the data points are connected with lines between them, the scatter diagram is referred to as a line graph. Connecting lines are not used so that they do not clutter the graph and obscure the trend line.

Although any two consecutive columns or rows or can be used for the data, when columns are used, the data entered in the left-most column become the x-axis values and the data in the right-most column become the y-axis values. For data listed in rows, the upper-most row data is plotted on the x-axis and the lowest-row data on the y-axis.

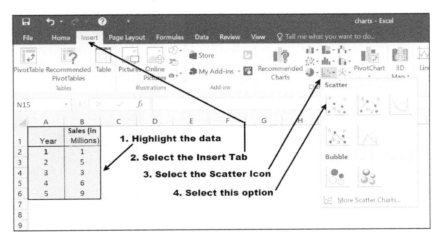

FIGURE 7.10 The steps required to produce a scatter diagram.

After selecting the scatter option listed as Step 4 in Figure 7.10, the scatter diagram shown in Figure 7.11 will appear. Notice that the figure has no axes titles and that the chart's title is a copy of the data label in cell B2, which identifies the y-axis values.

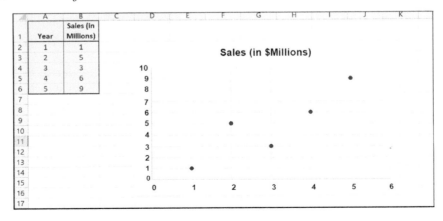

FIGURE 7.11 The initial scatter diagram.

Changing a Scatter Diagram's Elements and Appearance

Figure 7.12 shows the initial position and description of the descriptive elements associated with line graphs. These consist of a chart title, axes labels, and legend. When any of the four textual elements are present (title, axes labels, and legend) the text, font type, and size of each of these elements can be changed by clicking on the desired element and directly entering new text and font attributes.

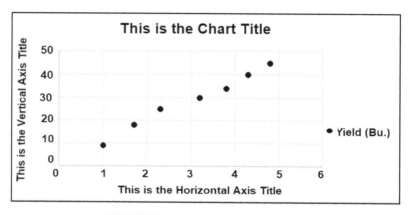

FIGURE 7.12 Basic scatter diagram elements.

Adding or deleting a scatter diagram's title, axes labels, legend, and grid lines is done using the Chart Elements submenu shown in Figure 7.13. This submenu is activated by either clicking on the Add-Chart Element in the Design ribbon or right or left-clicking within the chart itself (item 1 in the figure), and then selecting the cross-hairs icon (item 2 in the figure). Clicking on a chart element within the submenu adds the element to the chart, while clearing a checkbox removes the element.

A scatter diagram's colors can be changed using the Change Colors option in the Chart Tools Design ribbon, as shown at the upper left in Figure 7.14. The Design ribbon is activated either by double-clicking within the chart or single clicking and then selecting the Design tab under the Chart Tools option (at the top of Figure 7.14).

For example, checking the Axis Titles and Legend checkboxes shown in Figure 7.13 produces the chart shown in Figure 7.14a. Simply double clicking within each element allows you to change the titles to those of your own choosing, including font type and size. This was done to produce the titles shown in Figure 7.14b.

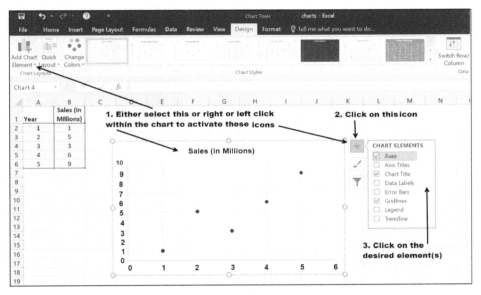

FIGURE 7.13 Adding or deleting chart elements.

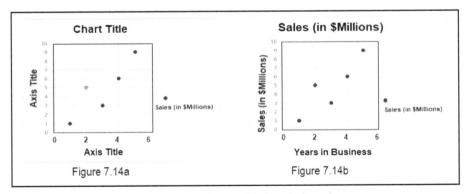

FIGURE 7.14 Adding axes labels and a legend.

A scatter diagram's appearance is changed by selecting one of the chart styles shown within the Chart Design ribbon (see Figure 7.15). Clicking on any of the pre-constructed styles changes the cart to the selected choice. Note that the chart elements previously shown in Figure 7.13 can also be accessed and changed by clicking on the leftmost icon in the Chart Design ribbon.

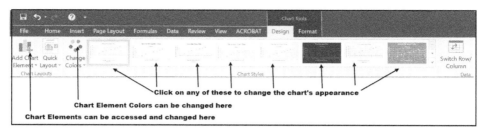

FIGURE 7.15 The chart design ribbon.

Adding a Trend Line

To add a trend line for the data shown on a scatter diagram, either click on the Add Chart Elements in the Chart Design Tools ribbon, as shown in Figure 7.16 (if this ribbon is not shown, double-clicking anywhere within the scatter diagram causes it to be displayed), or click on the Chart Elements Icon (if this Icon is not displayed, either right or left-clicking within the scatter diagram activates it). Either of these actions produces the submenus shown in Figure 7.16.

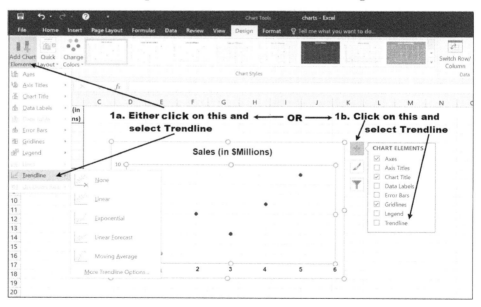

FIGURE 7.16 Adding chart elements.

Clicking on the arrow to the right of the Trendline option from either of the submenus displayed in Figure 7.16 (hover over the Trendline option to activate the arrow) triggers the menu shown in Figure 7.17.

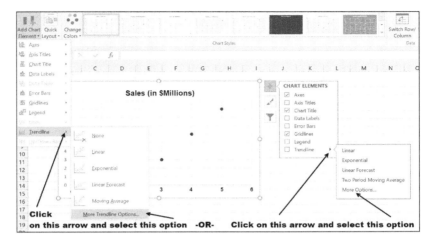

FIGURE 7.17 Selecting alternate trendline curves.

When the Trendline drop-down menu shown in Figure 7.17 is displayed, you are ready to add the Trend line and its equation onto the scatter diagram. To do so, click on the last option, which is labeled More Trendline Options. Selecting this option will bring up the last menu needed, which is that shown on Figure 7.18.[2]

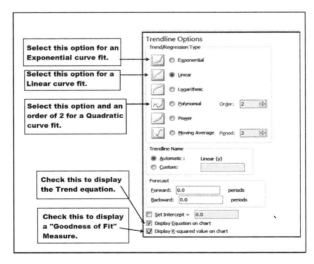

FIGURE 7.18 Trendline options.

---

[2] Note: Checking the Trendline box on the Chart Elements submenu immediately produces a linear trend with no equation.

When the menu shown in Figure 7.18 is displayed, click on the desired trend line type as shown in the figure. Then check the *Display Equation* and *R-squared value on Chart* checkboxes at the bottom of the menu and click on the Close button. This will produce a display of the chosen trend line with its associated equation and *R*-squared value directly on the scatter diagram. Selecting a Linear Trend line for the Year and Sales data used throughout this example produces Figure 7.19.

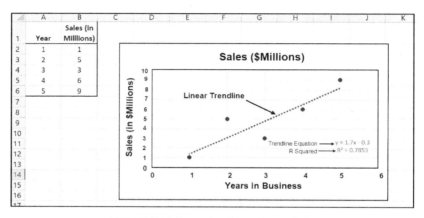

FIGURE 7.19 A completed trendline analysis.

Exercises 7.3

1. Construct a scatter diagram for the following data:

Year	Sales ($Millions)
1	0.5
2	1.7
3	1.9
4	1.4
5	1.6
6	2.0
7	2.2
8	2.4
9	2.9
10	3.3

2a. Construct a scatter diagram for the following sales data:

Year	1	2	3	4	5	6	7	8
Sales	$15,000	$18,000	$16,000	$20,000	$18,000	$22,000	$19,000	$24,000

b. Add appropriate chart and axes titles to the scatter diagram created for Exercise 2a.

3a. The share of people in the world with water available from a protected source is given in the following table (*Source: World Health Organization*). Construct a scatter diagram for this data. (**Hint**: Code Year 1980 as 1, 1990 as 10, and so on.)

Year	1980	1990	2000	2010	2015
Percentage	58%	68%	76%	84%	88%

b. Add appropriate chart and axes titles to the scatter diagram created for Exercise 3a.

4a. Construct an Excel scatter diagram and a linear trend line for the following data:

x	0	1	2	3	4	5	6
y	1	5	8	7	12	14	13

b. Add appropriate chart and axes titles to the scatter diagram created for Exercise 4a and make sure that the trend line's equation and R^2 value are shown on the chart.

5a. Construct an Excel scatter diagram and a linear trend line for the following data:

x	0	1	2	3	4	5	6	7
y	36	49	55	56	67	69	76	85

b. Add appropriate chart and axes titles to the scatter diagram created for Exercise 5a and make sure that the trend line's equation and R^2 value are shown on the chart.

6a. Construct an Excel scatter diagram and a linear trend line for the following data:

Year	2014	2015	2016	2017	2018	2019	2020
Sales ($Thousands)	15	18	16	20	18	22	19

b. Add appropriate chart and axes titles to the scatter diagram created for Exercise 6a and make sure that the trend line's equation and R^2 value are shown on the chart. (**Hint**: Code Year 2014 as 1, 2015 as 2, and so on.)

7a. Construct an Excel scatter diagram and a 2nd order polynomial (quadratic) trend line for the following data:

x	-2	-1	0	1	2
y	10	11	15	16	23

b. Add appropriate chart and axes titles to the scatter diagram created for Exercise 7a and make sure that the trend line's equation and R^2 value are Shown on the chart.

8a. Construct an Excel scatter diagram and a 2nd order polynomial (quadratic) trend line for the following data:

x	-3	-2	-1	0	1	2	3
y	8.1	2.9	0	-1	0	3.1	7.8

b. Add appropriate chart and axes titles to the scatter diagram created for Exercise 8a and make sure that the trend line's equation and R^2 value are shown on the chart.

9a. Construct an Excel scatter diagram and an exponential trend line for the following data:

x	0	1	2	3	4
y	2	16	90	500	2,500

b. Add appropriate chart and axes titles to the scatter diagram created for Exercise 9a, and make sure that the trend line's equation and R^2 value are shown on the chart.

10a. Construct an Excel scatter diagram and an exponential trend line for the following data:

x	0	1	2	3	4	5
y	16	20	38	60	90	130

b. Add appropriate chart and axes titles to the scatter diagram created for Exercise 10a, and make sure that the trend line's equation and R^2 value are shown on the chart.

11a. Construct an Excel scatter diagram and an exponential trend line for the following data:

Year	0	1	2	3	4
Sales ($1,000)	0.9	2.9	9.5	28.8	100

b. Add appropriate chart and axes titles to the scatter diagram created for Exercise 11a, and make sure that the trend line's equation and R^2 value are shown on the chart.

c. Use the trend equation found in Part a to project sales for the 6th year.

12a. Construct an Excel scatter diagram and an exponential trend line for the following Data (**Hint**: Code Year 2014 as 1, 2015 as 2, and so on):

Year	2016	2017	2018	2019	2020
Population	4,953	7,398	11,023	16,445	24,532

b. Add appropriate chart and axes titles to the scatter diagram created for Exercise 12a, and make sure that the trend line's equation and R^2 value are shown on the chart. Use the trend equation found in Part a to project the population in 2022.

13a. Using R^2 values, determine whether a linear, quadratic, or exponential trend line best fits the following data:

Years in Business	1	2	3	4	5
Sales ($Milions)	10	14	26	38	55

b. Use the trend line found to be best in Exercise 13a to project sales for the company's 6th and 7th year in business.

7.4 SELECTING AN APPROPRIATE TREND LINE USING R^2

The example in the prior section illustrated the modeling of the data using a linear trend line. In practice, the choice of which trend line to use is always based on the data itself. Thus, in practice, after a scatter diagram has been produced, the next step is to determine which, if any, of the curve types considered in the prior section best fits the data. The starting point is always the same: Use a scatter diagram to plot the data and then take a long, hard look at the resulting plot.

Typically, a pattern develops. Based on the emergent pattern and knowledge of various curves, a curve type that appears to fit the data reasonably well is then selected. As shown in Figure 5.41, Excel provides a number of Trendline options.

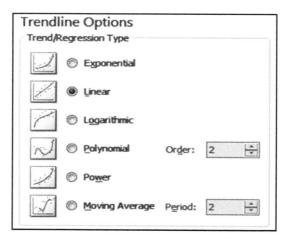

FIGURE 7.20 Excel's trendline options.

The final step is then to determine whether or not the resulting line does, in fact, adequately model the data. It is here that the R^2 value, which is formally known as the ***Coefficient of Determination***, comes into play.

Mathematically, an R^2 of 0 indicates that the variable plotted on the y-axis cannot be predicted from the value plotted on the x-axis. Similarly, an R^2 of 1 indicates that the variable on the y-axis can be predicted, without error, from the corresponding value on the x-axis, assuming all other variables remain the same. An R^2 between 0 and 1 indicates the degree to which the y-axis variable is predictable. Thus, an R^2 of 0.25 indicates that 25% of the y-axis value is predictable from x-axis value, and an R^2 of 0.80 means that 80% is predictable. In all cases, this requires that all other variables remain constant. Interestingly, the coefficient of determination, R^2, is the square of the ***Correlation Coefficient*** between y-axis values and x-axis values[3]. An R value of 1 indicates perfect correlation, while a value of 0 indicates no correlation. It is important to understand that ***even if two quantities are perfectly correlated,***

---

[3] Although the correlation coefficient squared yields the coefficient of determination, the square root of the coefficient only provides the *value* of the correlation coefficient, not whether it is positive of negative. In addition to its value, correlation coefficients have a sign, either + or –, indicating either a positive or negative correlation, respectively.

correlation does not imply causation. All that correlation implies is that when one quantity occurs, the other quantity tends to occur to the degree indicated by the correlation coefficient. *It never means* that one quantity causes the other quantity to occur.

7.5   SUMMARY OF KEY POINTS

Key Terms

- Coefficient of Determination
- Correlation Coefficient
- Linear least squares
- Scatter diagram
- Trend line

Key Concepts

- Curve fitting is a technique where actual data is used to determine an equation relating the variables under consideration.
- The simplest curve-fitting equation is known a constant curve fit. Here, the curve is a straight line with zero slope, with an equivalent equation of the form y = constant. The constant is the average of the independent variables.
- A Trend Line is a line that best fits the data following user-defined criteria.
- The most commonly used criteria for constructing a Trend Line is the linear least-squares method.
- The linear least-squares criteria minimizes the sum of the squared error between the Trend Line and the actual data.
- The most commonly used Trend Lines are typically either linear, polynomial, exponential or logarithmic curves.
- Trend Lines are easily determined using Excel by first creating a scatter diagram and then superimposing a Trend Line on the scatter diagram.
- The Coefficient of Determination is used to determine the degree of accuracy of a Trend Line.

SOLUTIONS TO SELECTED ODD-NUMBERED EXERCISES

CHAPTER 1

Section 1.1

1. -3 **3.** 1.6 **5.** $-9\ 1/2$ **7.** 162 **9.** -162 **11.** $-2/3$ **13.** -14.03

15. 4 **17.** -4 **19.** $-4/5$ **21.** -5.5 **23.** -4 **25.** 4 **27.** -4

29. -7.7 **31.** 4 **33.** 2 **35.** -6.62 **37.** 30 **39.** $246/497$

Section 1.2

1. 81 **3.** π^{15} **5.** $(1.7)^{5.1}$ **7.** y^{10} **9.** $(3.1)^{24}$

11. $5^4 = 625$ **13.** $(2.15)^3 = 9.26$ **15.** $\pi^3 = 31$ **17.** 27

19. 9 **21.** 6 **23.** 0.1 **25.** 1.5

27. 5.6 **29.** $x = 2$ **31.** $p = 2$ **33.** $w = 5$

Section 1.3

1. Yes **3.** No **5.** No **7.** No

9. Yes **11.** No **13.** Yes **15.** $x = 5$

17. $x = 2$ **19.** $P = 4$ **21.** $T = 7$ **23.** $x = -3$

25. $p = -23/6$ **27.** $a = 13/3 = 4.33$ **29.** $y = 1/3 = .33$ **31.** -2457

Section 1.4

1. a. $(x_1)^2 + (x_2)^2 + (x_3)^2$

b. $2x_3 + 2x_4 + 2x_5 + 2x_6 + 2x_7 + 2x_8 + 2x_9 + 2x_{10} + 2x_{11}$

c. $+ (x_1 + y_1) + (x_2 + y_2) + (x_3 + y_3) + (x_4 + y_4) + (x_5 + y_5) + (x_6 + y_6)$

d. $(3)M_{99} + 4) + (3)M_{100} + 4) + (3)M_{101} + 4) + (3)M_{102} + 4)$
$\qquad + (3)M_{103} + 4) + (3)M_{104} + 4) + (3)M_{105} + 4)$

3. a. $\displaystyle\sum_{i=2}^{29} 3i^2$ **b.** $\displaystyle\sum_{i=2}^{29} i(3^2)$ **c.** $\displaystyle\sum_{i=2}^{29} 2(3^i)$ **d.** $\displaystyle\sum_{i=2}^{29} (-1)^i(3i^2)$

5. a. 15   **b.** 31   **c.** 151   **d.** 3   **e.** 26   **f.** 465   **g.** They are not equal.

9. Average $= 1/n \displaystyle\sum_{i=1}^{n} G_i$

Section 1.5

1. 0.67   **3.** 0.24   **5.** 2.87

7. 0.667   0.235   2.871

9. 3356.42   **11.** .003356   **13.** 3,300,000,000,000

CHAPTER 2

Section 2.1

1. a.   A: $(3, 2)$ B: $(9, 6)$ C: $(10,0)$ D: $(4, -6)$ E: $(8, -4)$
  F: $(-6, 5)$ G: $(-2, 1)$ H: $(0, 5)$ I: $(-5, -3)$
  J: $(-1, -4)$ K: $(0, -7)$

b. Points A and B

3.

a.

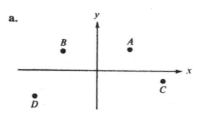

b.

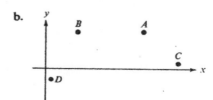

c.

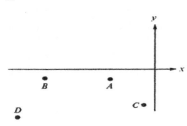

5.

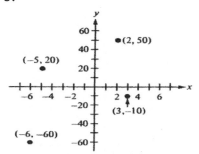

9. a. All points on the x-axis have a y-coordinate of 0.

   b. All points on the y-axis have an x-coordinate of 0.

11. Each point on the line has the same y-coordinate.

Section 2.2

1.

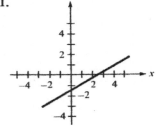

3.

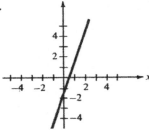

5.

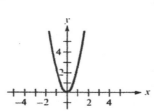

7.

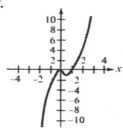

9.

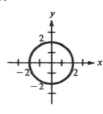

11. (a)

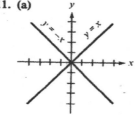

    b. The curves have opposite slopes.

  13. All three points lie on the same graph.

Section 2.3

  1. a. Yes   **b.** No   **c.** No   **d.** Yes   **e.** Yes   **f.** Yes

    g. No   **h.** Yes   **i.** Nos   **j.** Yes   **k.** Yes   **l.** Yes

  3.

    1. a.

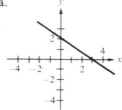

    b.

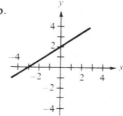

c.
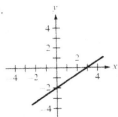

d.
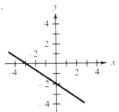

e.
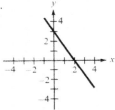

f.
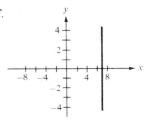

g.
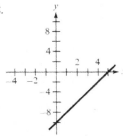

h.
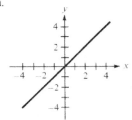

5. $V = \$6,000 - \$1,500t$ is a linear equation.

7. 23 months

Section 2.4

1. a. $-2/3$ **b.** $2/3$ **c.** $2/3$ **d.** $-2/3$

  e. $-3/2$ **f.** ∞ **g.** 2 **h.** 1

3. $A = (2/11)t + 1$

5. $P = 0.05N + 2000$

7. $P = 1.9x + 1$

9. a. $y = 250x + 35,000$ ←this is a linear equation

  b. $\$52,500$

  c. 160

Section 2.5

1. a. $C = 55x + 225,000$ **b.** $R = 100x$

    c. $45,000   **d.** 5,000 units

3. Break-even point = $15,000 / ($12.00 - $2.00) = 1,500 units

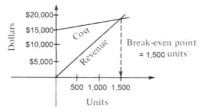

5. a. 9,375   **b.** 6,383

    c. Yes, the higher the price the greater profit on each unit. This means break-even will occur at a lower number of units sold.

7. The variable cost per bookend is $9.00

CHAPTER 3

Section 3.1

1. Yes   **3.** No   **5.** Yes   **7.** No

9. Yes   **11.** Yes   **13.** Yes

15. a. Yes   **b.** Yes   **c.** No

     d. Yes   **e.** No   **f.** No (consider two-tone cars)

     g. No

Section 3.2

1. a. 0, 1, 2, 3, 4, 5, 6, 7, 8, 9, 10

    b. Integer numbers

    c. 225, 196, 169, 144, 121, 100, 81, 64, 49, 36, 25

3. Yes; the inverse is also a function.

5. Yes, the inverse is not a function.

7. a. 4   **b.** 34   **c.** −6   **d.** $(a + b)^2 + 3(a + b) - 6$

    e. $(x + \Delta x)^2 + 3(x + \Delta x) - 6$

9. a. 18 **b.** $d + 2d^2 + d^3$ **c.** $(x + y) + 2(x + y)^2 + (x + y)^3$
d. $(2a) + 2(2a)^2 + (2a)^3 = 2a + 8a^2 + 8a^3$

Section 3.3

1. polynomial of degree 2
3. polynomial of degree 2
5. polynomial of degree 0
7. polynomial of degree 0
9. polynomial of degree 5
11. *not* a polynomial
13. *not* a polynomial
15. *not* a polynomial
17. *not* a polynomial
19. 11, 13, 16, 17, 18
21. the sum of the degrees of each polynomial in the product
23. If $f(x)$ is a polynomial of degree n, then $f(0) = a_n(0)^n + \ldots + a_2(0)^2 + a_1(0) + a_0$.
The point $(0, a_0)$ is the y-intercept.

Section 3.4

1. a. Yes, x **b.** No   **c.** Yes, x **d.** Yes, x
e. No   **f.** Yes, n **g.** Yes, S **h.** Yes, x
i. Yes, x **j.** Yes, y **k.** Yes, x **l.** No

3. a. $x = 3, -2$ **b.** $x = 5/3, -1$

c. $x = -\dfrac{\sqrt{7}}{2}, +\dfrac{\sqrt{7}}{2}$

d. $x = \dfrac{3 \pm \sqrt{21}}{2}$

e. $x = \dfrac{1 \pm \sqrt{17}}{2}$

f. $x = \dfrac{4 \pm \sqrt{8}}{2} = \dfrac{4 \pm \sqrt{4.2}}{2} = \dfrac{4 \pm 2\sqrt{2}}{2} = 2 \pm \sqrt{2}$

g. $x = -1, 2$ **h.** $x = 0, 1$ **i.** $x = 0, 4$

5. a. $15,125 **b.** $20,000 **c.** 6 years

7. a. 15 **b.** cost per cup > revenue per cup **c.** $165

Section 3.5

1. Yes, $a = 2, b = 7$

3. No

5. $7^{2x} = (7^2)^x = 49^x$, $1 = 1$, $b = 49$.

7. Yes, $a = \pi, b = e$

9. Yes, $a = 1, b = 7$

11. a. 1,897.39   **b.** 18,000.38

13. a. 88.69 grams   **b.** 235.44 grams

CHAPTER 4

Section 4.1

1. a. $240   **b.** $1,240

3. a. $7.50   **b.** $1.007.50

5. a. $843.75   **b.** $3343.75

7. a. $164   **b.** $3,364

9. a. $10,237.50   **b.** $17,062.50

Section 4.2

1. a. i = .02, $n = 4$, $P(0) = \$2,000$

   b. $2,164.86

3. $2,812.16

5. $2,901.89

7. $1,853.94

9. $3,033.40

11. $P(1460) = \$3,500\,(1 + .025/365)^{1460} = \$3,868.08$

13. $P(1095) = \$2,500\,(1 + .03/365)^{1095} = \$2,735.43$

Section 4.3

1. a. $7,401.22 **b.** $7,429.74 **c.** $7,444.32 **d.** $7,458.96

3. $16,991.90

5. At 2% compounded quarterly, $1,000 is worth $1,020.15 after a year.
At 4% compounded annually, $1,000 is worth $1,040 after a year.
Therefore, the 4% annually is the better investment alternative.

7. a. The savings plan yields $2091.36

  b. The loan yields $2,283.92

  c. Not taking into account credit risk, the loan is the better investment

9. a. $12,802.36 **b.** $10,960.35 **c.** $8,104.03

  d. The Present value of a Future Value amount goes down (that is, decreases) as the interest rate increases.

11. $8,367.55

13. In the friend's business she will receive $12,000 in three years.
Investing it in the 4% account yield $11,268.25. Therefore, assuming equal credit risk for both alternative, she gets a better return from her friend's business.

15. *PV* of Buyer A = $20,000 + $4,274.02 = $24,274.02

  PV of Buyer B = $15,000 + $8,889.96 = $23,889.96

  PV of Buyer C = $10,000 + $14,225.66 = $24,225.66

  Assuming equal credit risk, Buyer A has made the best offer.

17. 1st Opportunity: *PV* = $6,336.75

  2nd Opportunity: *PV* = $6,229.98

  3rd Opportunity: *PV* = $6,650.57

  Assuming equal credit risk, the 3rd opportunity is better.

Section 4.4

1. *NPV* = $735.29 + 1,057.29 + $1.847.69 − $2,500

    = $3,640.27 − $2,500 = $1,140.27

3. *NPV* = 493.80 + $969.26 + $1,855.63 − $2,900 = $418.69

5. *NPV* = $778.59 + $1,420.78 + 3,226.42 − $5,000 = $425.80

7. *NPV* = $577.35 + $1,048.79 +$1,500.28 = $3134.42

Section 4.5

1. $$PV = \$50\left[\frac{1-\left(1+\dfrac{0.04}{4}\right)^{-40}}{\dfrac{0.04}{4}}\right] = \$1,641.73$$

3. $$\text{Net } PV = \$750\left[\frac{1-(1+0.08)^{-3}}{0.08}\right] - \$1,500 = \$1,932.82 - \$1,500 = \$432.82$$

5. $$NPV_1 = \$2,000\left[\frac{1-\left(1+\dfrac{0.04}{12}\right)^{-24}}{\dfrac{0.04}{12}}\right] + \$38,000\left(1+\dfrac{.04}{12}\right)^{-24} - \$70,000$$

$$= \$46,056.50 + \$35,083.09 - \$70,000 = 11,139.59$$

$$NPV_2 = \$3,500\left[\frac{1-\left(1+\dfrac{0.04}{12}\right)^{-24}}{\dfrac{0.04}{12}}\right] - \$70,000$$

$$= \$80,598.88 - \$70,000 = \$10,598.88$$

On a strictly monetary basis, and assuming equal credit risk for both investments, the first investment is more profitable.

7. $$NPV_1 = \$2,300\left[\frac{1-\left(1+\dfrac{0.04}{4}\right)^{-12}}{\dfrac{0.04}{4}}\right] - \$20,000$$

$$= \$25,886.68 - \$20,000 = 5,886.68$$

$$NPV_2 = \$1,500\left[\frac{1-\left(1+\dfrac{0.04}{4}\right)^{-16}}{\dfrac{0.04}{4}}\right] + \$2,000 - \$18,000$$

$$= \$22,076.81 + \$2,000 - \$18,000 = \$6,076.81$$

On a strictly monetary basis, and assuming equal credit risk for both investments, the second investment is more profitable (Note: to compare investments the same amount must be considered in both NPVs. Thus, in the second alternative, after investing $18,000, there is still $2,000 of immediate, or Present Value, funds available, which must be added to the NPV).

9. $PV = \$80\left[\dfrac{1-\left(1+\dfrac{0.06}{2}\right)^{-20}}{\dfrac{0.06}{2}}\right] + \$1,000\left(1+\dfrac{.06}{2}\right)^{-20}$

$= \$1,190.20 + \$553.68 = \$1,743.88$

11. $FV = \$400\left[\dfrac{\left(1+\dfrac{0.04}{4}\right)^{12}-1}{\dfrac{0.04}{4}}\right] = \$400\left(12.682503\right) = \$5,073.00$

13. $FV = \$20\left[\dfrac{\left(1+\dfrac{0.05}{52}\right)^{48}-1}{\dfrac{0.05}{52}}\right] = \$20\left(49.100757\right) = \$982.02$

15. $PV = \$750\left[\dfrac{1-\left(1+\dfrac{0.075}{12}\right)^{12}}{\dfrac{0.075}{12}}\right] + \left(\$1,000\right)\left(1+\dfrac{.075}{12}\right)^{-12}$

$= 864.40 + \$925.96 = \$1,792.44$

Section 4.6

1. $171.87

3. $(\$171.87)(360) - \$36,000 = \$25,873.20$

7. a. $1,016.06

b. $(\$1,016.06)(48) - \$45,000 = \$48,770.88 - \$45,000 = \$3,770.88$

c.

	A	B	C	D	E	F	G
1	Amount of Loan:		$45,000				
2	Length of Loan (in years):		4				
3	Annual Interest Rate:		4%				
4	Montly Payment		$1,016.06				
5							
6			**Payment Number**	**Payment Number**	**Interest Paid**	**Principal Paid**	**Outstanding Balance**
7			0	-	-	-	$45,000.00
8			1	$1,016.06	$150.00	$866.06	$44,133.94
9			2	$1,016.06	$147.11	$868.95	$43,265.00
10			3	$1,016.06	$144.22	$871.84	$42,393.16

9. $PMT = \$295.60$ Interest $= (\$295.60)(300) - (\$40,000) = \$48,680$

Section 4.7

1. a. $(\$480 + \$3,000) / 24 = \$145.00$

  b. $(\$720 + \$3,000) / 24 = \$103.33$

  c. 14.677% for the 2 year loan, 14.546% for the 3 year loan

3. a. Total Interest $= (0.06)(\$4,000)(1\text{year}) = \$240.00$

  b. Monthly payment $= (\$240 + \$4,000) / (12 \text{ months}) = \$353.33$

  c. For a PV of $4,000, a payment of $353.33 for 12 months (1 year), the true interest rate is 10.895%.

5. a. Monthly payment $= \$9,000/36 \text{ months} = \$250$

  b. Total Interest charge $= (0.06)(\$9,000)(3 \text{ years}) = \$1,620$

  c. Cash Received $= \$9,000 - \$1,620 = \$7,380$

  d. For a PV of $7,380, a payment of $250 for 36 months, the true interest rate is 13.376%.

7. a. Required loan amount $= (\$15,000) / [1 - (0.04)(5)] = \$18,750$

  b. Payment $= \$18,750/60 = \$312.50$

  c. Total interest $= \$18,750 - \$15,000 = \$3,750$

  d. For a PV of $15,000, a payment of $312.50 for 60 months (5 years), the true interest rate is 9.154%.

Section 4.8

1. $PV = \$1,000 + \$1,000 \left[\dfrac{1 - (1 + 0.04)^{-9}}{0.04} \right]$

$= \$1,000 + \$1,000(7.435332) = \$8,435.33$

3. $PV = \$20 + \$20 \left[\dfrac{1 - \left(1 + \dfrac{0.02}{12}\right)^{-179}}{\dfrac{0.02}{12}} \right]$

$= \$20 + \$20[154.6570781] = \$3,113.14$

5. $FV = \$40 \left(1 + \dfrac{0.04}{4}\right) \left[\dfrac{\left(1 + \dfrac{0.04}{4}\right)^{12} - 1}{\dfrac{0.04}{4}} \right]$

$= \$40(1.01)[12.682503] = \$512.37$

7. $FV = \$20 \left(1 + \dfrac{0.05}{52}\right) \left[\dfrac{\left(1 + \dfrac{0.05}{52}\right)^{48} - 1}{\dfrac{0.05}{52}} \right]$

$= \$20(1.000961541)[49.10075664] = \$982.96$

9. $FV = \$1 \left(1 + \dfrac{0.05}{365}\right) \left[\dfrac{\left(1 + \dfrac{0.05}{365}\right)^{730} - 1}{\dfrac{0.05}{365}} \right]$

$= \$(1.000961541)[767.6906738] = \$767.80$

11. $FV = \$1,000(1 + 0.08) \left[\dfrac{(1 + 0.08)^7 - 1}{0.08} \right]$

$= \$1.000(1.08)[8.92280335] = \$9,636.63$

Section 4.9

1. 2.0150%
3. 2.0201%
5. 4.0811%
7. 8.2432%

CHAPTER 5

Section 5.1

1. Average rate of change = 5.
3. Average rate of change = 16
5. Average rate of change = 11
7. Average rate of change = 3
9. Average rate of change = 31

Section 5.2

1. Instantaneous rate of change at $(1, -2)$ = slope of the tangent line = 3
3. Instantaneous rate of change at $(1, 1)$ = slope of the tangent line = -2
9. **a.** The average rate of change for all the given intervals is 0
   b. From the left the instantaneous rate of change is zero.
   From the right the instantaneous rate of change is zero.
11. **a.** $[\{3(7) + (1/2)(7)^2\} - 0]/7 = 6.5$ million per year
    b. \$10 million per year   **c.** \$80 million   **d.** \$75.5 million
13. **a.** The derivative is positive from $x = -12$ to $x = -6$, $x = 1$ to $x = 6$, and $x = 9$ to $x = 10$.
    b. The derivative is negative form $x = -6$ to x = 1, and x = 6 to x = 9.
    c. The derivative is zero at $x = -6$, $x = 1$, $x = 6$ and $x = 10$ to $x = 13$
15. **a.** No
    b. Yes, t The equation $D = 300p^2 + 50p + 200$ is the only equation where D increases as p increases.
    c. No
    d. No

Section 5.3

1. derivative = −3 $\qquad$ −3 and −3

3. derivative = $2x - 2$ $\qquad$ 0 and −12

5. derivative = $6x^2 - 4x + 3$ $\qquad$ 5 and 173

7. derivative = $2/(x + 2)^2$ $\qquad$ 2/9 and 2/9

9. derivative = $-2/x^3$ $\qquad$ −2 and 2/125

11.

x	Limit from the Left	Limit from the Right	Limit	Functional Value	Continuous at x
1	−1	−2	None	−1	No
2	1	2	None	1	No
3	2	−1	None	2	No

13.

x	Limit from the Left	Limit from the Right	Limit	Functional Value	Continuous at x
1	1	1	1	−1	No
2	1	1	1	−1	No
3	1	1	1	1	Yes

Section 5.4

1. $f'(x) = 0$ $\qquad$ **3.** $f'(x) = 0$ $\qquad$ **5.** $f'(x) = 2^2$

7. $f'(x) = 4x^3$ $\qquad$ **9.** $f'(x) = 16x^{15}$ $\qquad$ **11.** $f'(x) = 118\,x^{117}$

13. $f'(x) = -x^{-3}$ $\qquad$ **15.** $f'(x) = -4x^{-5}$ $\qquad$ **17.** $f'(x) = -16x{-1}$

19. $f'(x) = -236x{-237}$ $\qquad$ **21.** $f'(x) = (\frac{1}{2})x^{-1/2}$ $\qquad$ **23.** $f'(x) = (1/3)^{-2/3}$

25. $f(x) = x^2$ $\qquad$ $f(3) = (3)^2 = 9;$

$\qquad$ $f'(x) = 2x$ $\qquad$ $f'(3) = 2(3) = 6$

27. $f(x) = 2x^4$ $\qquad$ $f'(3) = 2(3)^4 = 162;$

$\qquad$ $f'(x) = 8x^3$ $\qquad$ $f'(3) = 8(3)^3 = 216$

29. $f'(x) = 7x$ $\qquad$ $f(3) = 7(3) = 21;$

$\qquad$ $f'(x) = 7$ $\qquad$ $f'(3) = 7'$

31. $f'(x) = e^x$

33. $f'(x) = 15e^x$

35. $f'(x) = 5e^5$

37. $f'(x) = 3e^{3x}$

39. $f(x) = -7e^{-7x}$

41. $f(x) = -2e^{-2x}$

43. $f(x) = 14e^{-7x}$

45. $f(x) = 10e^{-2x}$

47. $dy/dx = (-1/2)x^{-3/2}$

49. $dy/dx = (-3/8) x^{-11/8}$

51. $dy/dx = 16x^7$

53. $dy/dx = 15x^2$

55. $dy/dx = 18x^5$

57. $dy/dx = 22x$

59. $dy/dx = 7$

61. $dy/dx = 5$

63. $dy/dx = -6x^{-3}$

65. $dy/dx = -8x^{-5}$

Section 5.5

1. $f'(x) = 15x^4 + 8x$

3. $f'(x) = 18x^2 + 4x + 14$

5. $f'(x) = 40^3 + 10x + 7$

7. $f'(x) = 15^4 - 8x$

9. $f'(x) = 12x^2 + 4x - 14$

11. $dy/dx = -40x^{-5} - 12x^{-3} + 3$

13. $dy/dx = -15x^{-6} + 8x^{-3}$

15. $dy/dx = -18x^{-4} - 4x^{-3} - 14$

17. $dy/dx = -15x^{-6} + 12x^{-4}$

19. $f'(x) = 15x^4 + 4e^x$

21. $f'(x) = 18e^{3x} + 14e^{2x} + 14$

23. $f'(x) = 30e^{6x} + e^x$

25. $f'(x) = -6x - 4e^{-2x}$

27. $f(x) = -40e^{-4x} + 15e^{3x} + 7$

29. $f(x) = 15e^{5x} - 4x^{-3}$

31. $f(x) = 18e^{3x} - 4e^{-2x} - 14$

33. a. $dS/dx = 6000 - 100x$

   b. Yes

35. a. 2 miles per 4 minutes = 30 miles per hour

b.

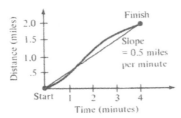

c. 0.65 miles per minute = 39 miles per hour

d.

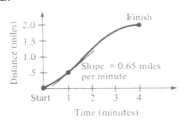

Section 5.6

1. $f'(x) = (8x + 2)(18x^2 + 36x^{-4}) + (6x^3 - 12x^{-3})\,8$

3. $f'(x) = (x^5 + 3)e^x + e^x(5x^4)$

5. $f'(x) = 9e^x(2x) + (x^2)(9e)^x$

7. $f'(x) = e^x(5x^4) + (x^5)\,e^x$

9. $f'(x) = (x^4)(18x^{-4} - 4e^{-2x}) + (6x^{-3} + 2e^{-2x})(4x^3)$

11. $f'(x) = 3x^7(7e^{7x}) + (e^{7x})(21x^6)$

13. $f'(x) = \dfrac{(x^5 + 3)e^x - e^x(5x^4)}{(x^5 + 3)^2}$

15. $f'(x) = \dfrac{(6x^{-3} + 2e^{-2x})(4x^3) - x^4(-18x^{-4} - 4e^{-2x})}{(6x^{-3} + 2e^{-2x})^2}$

17. $f'(x) = \dfrac{(e^{4x})3 - 3x(4e^{4x})}{(e^{4x})^2}$

19. $f'(x) = \dfrac{(x^4)(-12x^{-5} - 4e^{-2x}) - (3x^{-4} + 2e^{-2x})(4x^3)}{x^8}$

21. $f'(x) = \dfrac{(14x^3)(15e^{3x}) - (5e^{3x})(42x^2)}{(196x^6)}$

23. Rewrite this as $f(x) = 10x^{-2} - 3x$; then $f'(x) = -20x^{-3} - 3$

25. Rewrite this as $f(x) = (1/2)x^{10} + (1/2)e^x$; then $f'(x) = 5x^9 + (1/2)e^x$

Section 5.7

1. $f'(x) = 3(6x^{-3} + 2e^{-2x})^2(-18x^{-4} - 4e^{-2})$

3. $f'(x) = 2(6e^{-4x} + 7x)(-24e^{-4x} + 7)$

5. $f'(x) = 8(2e^{9x} - 6x^{-2})^7(18e^{9x} + 12x^{-3})$

7. $f'(x) = 15(10x^3 + 2e^{-2x})^{14}(30x^2 - 4e^{-2x})$

9. $dS/dx = (dS/dE)(dE/dx) = (2E + 3)(2x) = (2x^2 + 3)(2x) = 4x^3 + 6x$

11. $dS/dx = (dS/dE)(dE/dx) = \{[(E + 2) - E]/(E + 2)^2\}\,(6x)$
$= \{2/(3x^2 + 5 + 2)^2\}\,(6x) = 12x/(3x^2 + 7)^2$

Section 5.8

1. $y' = 5x^4 + 8x + 3$ $y'' = 20x^3 + 8$ 28 and 168

3. $y' = 4x^3 + 6x + 4$ $y'' = 12x^2 + 6$ 18 and 54

5. $y' = (10x^3 + 3x^2)e^{10X}$ $y'' = (100x^3 + 60x^2 + 6x)e^{10X}$ $166\,e^{10}$ and $1052\,e^{20}$

7. a. $-0.15t^2 + 0.5t + 0.3$

    b. 0.65 mile per minute = 39 miles per hour

    c. $-0.3t + 0.5$

    d. 0.2 mile per minute^2, -0.4 mile per minute^2 at $t = 3$ the car is decelerating or slowing down

CHAPTER 6

Section 6.1

1.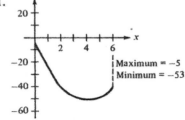

3.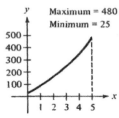

5.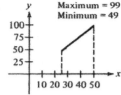

7. Maximum $= 23$; minimum $= -9$

9. Maximum $= 7$; minimum $= -9$

11. Maximum $= 559$; Minimum $= -305$

13. Maximum $= 12$; minimum $= -220/3$

15. Maximum $= 75$; minimum $= -25$

17. a. 22,000   **b.** \$232,000

19. a. 100 units   **b.** \$5000   **c.** \$25,000

21. 45 units

23. Maximum $= 6$; Minimum $= 0$

25. 5, 3, 2, 3.5

27. No; No

29. Relative minimum at $x = 10$, relative maximum at $x = 4$

Section 6.3

1. a. $P = -1/2x^2 + 180x - 5,000$ **b.** $0 \le x \le 300$ **c.** \$11,200

3. a. $P = 10x - (x^2/4,000 - 5x + 50,000) = -x^2/4,000 + 15x - 50,000$

b. $0 \le x \le 50,000$

c. \$175,000

5. 10,000 units

7. a. $TR = 500x - 2x^2$ **b.** $P = -2x^2 + 200x - 2,000$

  c. $0 \le x \le 40$ **d.** 40 units

9. a. 15 **b.** $200 - x/2$ **c.** $2.25 - .00005x$

11. a. $MC = x/2500 + 8$ **b.** $MC = 50$ **c.** $MC = 0.25$

13. $R = 200x$, therefore, $MR = R' = 200$

  $C = 5000 + 20x + (1/2) x^2$ therefore, $MC = C' = 20 + x$

  Setting $MR = MC$ yields $x = 180$

15. $R = 500x - 2x^2$, therefore, $MR = R' = 500 - 4x$

  $C = 300x + 2,000$, therefore, $MC = C' = 300$ Setting $MR = MC$ yields $x = 50$

Section 6.4

1. 60 **3.** 20 **5. a.** 100 **b.** 10 **7. a.** 800 **b.** 3 **c.** 2,400

9. Reorder point $= (31, 250/250) (2) = 250$

  Cycle time $= (250/2) = 10$ days

Section 6.5

1. $62.5 million   **3.** $62.5 million   **5.** $2,383.36 increase

CHAPTER 7

Section 7.1

1. $y = 10.3$

3. $y = 28.67$

5. An increase does not seem advisable, as sales appear to decrease as advertising is increased. (In the next section, the average rate of change of sales with respect to advertising can be determined, which will show that sales do, in fact, decrease with an increase in advertising – see the solution to Exercise 1 in Section 6.2.)

Section 7.3

1.

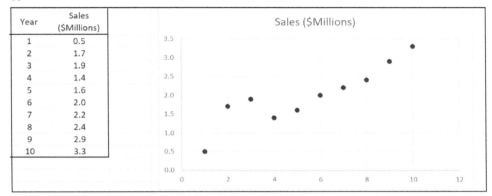

Year	Sales ($Millions)
1	0.5
2	1.7
3	1.9
4	1.4
5	1.6
6	2.0
7	2.2
8	2.4
9	2.9
10	3.3

3. b.

Year	0	10	20	30	40
Percentage	58%	68%	76%	84%	88%

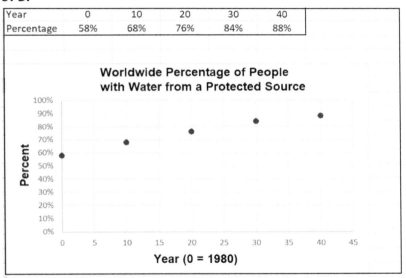

5. b.

x	0	1	2	3	4	5	6	7
y	36	49	55	56	67	69	76	85

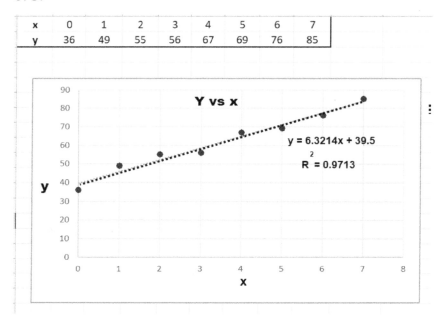

7. a. $y = .6429x^2 + 3.1x + 13.714$ $R^2 = .9612$

9. b. $y = 2.633e^{1.7704x}$ $R^2 = .9991$

11. b. $y = .8971e^{1.717x}$ $R^2 = .9995$

  c. Projected 6th year sales = \$1,104.41

13. a. A quadratic trend line has the highest $R^2 = .9975$

  b. The quadratic trend line is $y = 1.8571x^2 + .257x + 7.4$

  Projected sales when $x = 6$ is \$75.80 million

  Projected sales when $x = 7$ is \$100.20 million

INDEX